16th International Conference on Metrology and Properties of Engineering Surfaces (Met & Props 2017)

Journal of Physics: Conference Series
Volume 1183

Gothenburg, Sweden
26 – 29 June 2017

ISBN: 978-1-5108-8939-2
ISSN: 1742-6588

Printed from e-media with permission by:

Curran Associates, Inc.
57 Morehouse Lane
Red Hook, NY 12571

Some format issues inherent in the e-media version may also appear in this print version.

Institute of Physics
Dirac House, Temple Back
Bristol BS1 6BE UK

Phone: 44 1 17 929 7481
Fax: 44 1 17 920 0979

techtracking@iop.org

Additional copies of this publication are available from:

Curran Associates, Inc.
57 Morehouse Lane
Red Hook, NY 12571 USA
Phone: 845-758-0400
Fax: 845-758-2633
Email: curran@proceedings.com
Web: www.proceedings.com

Table of contents

Volume 1183

16th International Conference on Metrology and Properties of Engineering Surfaces (Met and Props 2017)

26–29 June 2017, Gothenburg, Sweden

Accepted papers received: 1 February 2019
Published online: 11 June 2019

Preface

Papers

Surface Properties and Function

Surface Metrology

Surface Manufacturing and Control

IOP Conf. Series: Journal of Physics: Conf. Series **1183** (2019) 011001 doi:10.1088/1742-6596/1183/1/011001

Preface

Proceedings of the 16th International Conference, Gothenburg, Sweden, 26th– 29th June, 2017

Organized by:

The International Program Committee of The International Conference Metrology and Properties of Engineering Surfaces, Met&Props

Halmstad University, Sweden

IOP Conf. Series: Journal of Physics: Conf. Series **1183** (2019) 011001 doi:10.1088/1742-6596/1183/1/011001

Greetings from Chairman of the Local Organizing Committee

On behalf of the International Programme Committee, I am proud to acknowledge and thank the hard work of the members of the Local Organizing- and International Scientific Committees in preparations and implementations for the successful 16'th conference held in Gothenburg and for the third time in Sweden (1997 in Gothenburg and 2003 in Halmstad). It was my privilege and pleasure to welcome you all to the 16th International Conference on Metrology and Properties of Engineering Surfaces in Gothenburg at Hotel Clarion Post.

The conference 2017 resulted in four successful days of: 121 delegates, 9 exhibitors, participants from 17 countries and 5 continents, 7 invited speakers, 59 oral presentations, and 28 posters.

Professor Tom Tomas and Professor Ken Stout who founded this series of conferences in the United Kingdom more than thirty years ago, the conference has not only produced a large amount of scientific publications over the years but have also inspired to the shaping of a strong international network of researchers and industry, ISO standardization, industrial usage of advanced surface metrology, scientific journal and e-journal, books, educations, joint research projects, "spring off" companies, patents and more.

The conference series have also strongly contributed to the establishing of an increased understanding world-wide of the value and importance of knowledge, tools and methods in association with metrology and properties of surfaces.

In this book of proceedings, we publish a selection of the full-paper presentations made at the conference in Gothenburg. The selection consist of 13 grouped into three sessions: *Surface properties and Function*, *Surface Metrology* and *Surface Manufacturing and Control*, based on their original sessions at the conference in Gothenburg.

We have here an excellent selection of papers from many of the world's experts, expressing the state-of-the art in their fields of research as well as their own ambition and results.

To complete the picture from the Gothenburg conference, also a special issue of the IOP e-journal Surface Topography, Metrology and Properties, STMP, will cover an additional selection of papers to be found at: http://iopscience.iop.org/journal/2051-672X

Again, On behalf of the International Programme Committee, I thank all who made the conference possible and warmly welcome you all back to Lyon, www.metprops2019.org in early July 2019.

Bengt-Göran, BG, Rosén
Halmstad, 9th of November 2018

IOP Conf. Series: Journal of Physics: Conf. Series **1183** (2019) 011001 doi:10.1088/1742-6596/1183/1/011001

International Programme Committee

Prof. Liam Blunt (University of Huddersfield, UK)
Prof. Mohamed El Mansori (Ecole Nationale Supérieure d'Arts et Métiers (ENSAM), France)
Prof. Bengt-Göran, BG, Rosén (Halmstad University, Sweden)
Prof. Tom R Thomas (Halmstad University, Sweden)
Prof. Hassan Zahouani University of Lyon - ENISE - Ecole Centrale de Lyon, France

International Scientific Committee

A. Archenti, Royal Institute of Technology, Sweden
J. Berglund, Swerea-IVF, Gothenburg, Sweden
F. Blateyron, Digital Surf, France
L. Blunt, University of Huddersfield, UK
D. Butler, Nanyang Tech. Uni. NTU, Singapore
C. Brown, Worcester Polytechic Institute, USA
L. Bååth, Halmstad University, Sweden
Z. Dimkovski, Halmstad University, Sweden
M. El-Mansori, Ecole Nationale Superieure d'Arts et Metiers , France
L. Eriksson, Jönköping University, Sweden
C. Evans, University of North Carolina - Charlotte, USA
C. Giusca, National Physical Laboratory, UK
S. Gröger, Chemnitz University of Technology, Germany
H. Haitjema, Mitutoyo Netherlands
W. Hongjun, Beijing Information Science & Technology University, China
M. Kalin, University of Lubljana, Slovenia
P. Krajnik, Chalmers University of Technology, Sweden
R. Leach, University of Nottingham, UK
R. Larsson, Luleå Unversity, Sweden
T.-Y. (Victor) Lin, Center for Measurement Standards, ITRI, Taiwan
M. Malburg, Digital Metrology, USA
P. H. Nilsson, Volvo Powertrain, Sweden
H. Norrgaard Hansen, Danish University of Technology DTU, Denmark
W. Hongjun, Beijing Information Science & Technology University, China
R. Ohlsson, Volva Powetrain, Sweden
U. Olofsson, KTH Royal Institute of Technology, Sweden
P. Pawlus, Rzeszow University of Technology, Poland
S. K. Ramasamy, USA
B.-G. Rosén, Halmstad University, Sweden
E. Schulz-Kornas, Max Planck Weizmann Center for Integrative Archaeology and Anthropology, Germany
P. Scott, University of Huddersfield, UK
J. Seewig, University of Kaiserslauten, Germany
S. Schütte, Linköping University, Sweden
J.-E Ståhl, Lund University, Sweden
T.R. Thomas, Halmstad University, Sweden
T. Vorburger, National Institute of Standards and Technology, USA
A. Wennerberg, Malmö University, Faculty of Odontology, Sweden
J. Westberg, Metrology West, Sweden
M. Wieczorowski, University of Poznan, Poland
X. Q. (Jane) Jiang, University of Huddersfield, UK
H. Zahouani, University of Lyon - ENISE - Ecole Centrale de Lyon, France

Local Organizing Committee

Prof. Bengt-Göran, BG, Rosén (Halmstad University, Sweden) – Chair
Prof. Lars Eriksson (Jönköping University, Sweden)
Assoc. Prof. Zlate Dimkovski (Halmstad University, Sweden)
Ass. Prof. Sabina Rebeggiani (Halmstad University, Sweden)
Dr. Cecilia Anderberg (Chalmers Univeristy of Technology and Getinge AB, Sweden)
Dr. Johan Berglund (Swerea-IVF, Gothenburg, Sweden)
M.Sc. Olena Flys (SP, Technical Research Institute of Sweden)

IOP Conf. Series: Journal of Physics: Conf. Series **1183** (2019) 011001 doi:10.1088/1742-6596/1183/1/011001

Session: Surface Properties and Function

Sub-session	Title	Authors	Organization
Properties and Function	Calculation of plasticity index of honed cylinder liner textures	P Pawlus , W A Grabon, D Czach	Rzeszow University of Technology, Poland
	Wear Analysis of Pin and Roller Surfaces	Z El-Ghoul[1], S John[1], Z Dimkovski[1], P-J Lööf[1], J Lundmark[2], J Mohlin[3]	[1] Halmstad University, Halmstad, Sweden; [2]Applied Nano Surfaces AB, Uppsala, Sweden; [3]Gnutti Carlo Sweden AB, Gothenburg, Sweden
	Research of the influence of the 2D and 3D surface roughness parameters of bearing raceways on the vibration level	S Adamczak, P Zmarzły	Kielce University of Technology, Faculty of Mechatronics and Mechanical Engineering, Department of Manufacturing Engineering and Metrology, Poland
Additive Manufacturing	Evaluation of the influence of selected parameters of Selective Laser Sintering technology on surface topography	Cz Kundera, T Kozior	Kielce University of Technology, Poland
Function: Biotribology	Focus variation measurement and advanced analysis of volumetric loss at the femoral head taper interface of retrieved modular replacement hips in replica	Karl Walton1, Petrucci Maila[2] Racasan Radu[1], Liam Blunt[1], Alister Hart [3], Paul Bills[1]	[1]EPSRC Centre for Innovative Manufacturing in Advanced Metrology, University of Huddersfield, UK; [2]Università di Bologna; [3]London Implant Retrieval Centre, University College London, UK.
Function: Physical Surfaces	Surface Topography of Nitrided Steel Surfaces	J Berglund[1,2], E Troell[1]	[1]1Department of Manufacturing, Swerea IVF, Mölndal, Sweden, [2]Department of Industrial and Materials Science, Chalmers University of Technology, Sweden.
Multiscale aspects of Engineering Surfaces II	Characterization of subsurface deformation of turned brasses: lead brass (CuZn39Pb3) and lead free brass (CuZn21Si3P)	Vijeth V Reddy[1], Pui Lam Tam[2], Amogh Vedantha Krishna[1] and B-G Rosén[1]	[1]Functional Surfaces Research Group, Halmstad University, Halmstad, Sweden [2]Department of Industrial and Materials Science, Chalmers University of Technology, Gothenburg, Sweden

Session: Surface Metrology

Sub-session	Title	Authors	Organization
Metrology: Characterization	Identification of the minimum chip thickness based on the two-dimensional wavelet analysis	W Makieła, D Gogolewski, Ł Nowakowski	Kielce University of Technology, Faculty of Mechatronics and Mechanical Engineering, Department of Manufacturing Engineering and Metrology, Kielce, Poland
Standardization	Uncertainty evaluation for absolute flatness measurement on horizontally aligned fizeau interferometer	Kitti Wirotrattanaphaphisan[1,3], Jariya Buajarern[2] and Suthep Butdee[1], T Benjurlux[3]	[1]Department of Production Engineering, Faculty of Engineering, King Mongkut's University of Technology North Bangkok [2]Department of Dimensional metrology, National Institute of Metrology, Thailand, [3]Department of Industrial Engineering, Faculty of Engineering, Rajamangala University of Technology Lanna Tak, Thailand.
Metrology: Instrumentation	Fast Measurement Method of Gear Parameters Based on Convexity Defect	Chen Guo-hua, Liu Xiao-long, Zhang Ai-jun, Wang Jun-yi	Beijing University of Chemical Technology, China
In-Line Metrology	Transparent Object Detection and Location Based on RGB-D Camera	Chen Guo-hua, Wang Jun-Yi , Zhang Ai-jun, Xing Jian, LI Bin	Beijing University of Chemical Technology, College of Mechanical and Electronic Engineering, China

Session: Surface Manufacturing and Control

Sub-session	Title	Authors	Organization
Control and Condition monitoring	Evaluation method of running performance for five-axis machining center based on the "S" specimen	H J Wang[1,2], F X Han[1], Y H Gu[2], B-G Rosén[2,3] and A N Zou[1,2]	School of Mechanical and Electrical Engineering, Beijing Information Science and Technology University, China, [2]Key Laboratory of Modern Measurement and Control Technology, Ministry of Education, Beijing Information Science and Technology University, China.[3]The Rydberg laboratory, Halmstad University, Halmstad, Sweden.
	Dynamic characteristics of the annular rotor components in a rotor-bearing system under high frequency stress conditions	Chenglin Wang[1], Huating Chen[2], Weichen Sun[2], Hongjun Wang[3]	[1]Beijing Wuzi University, China, [2]Beijing Institute of Technology, China, [3]Beijing University of Technology, China.

IOP Conf. Series: Journal of Physics: Conf. Series **1183** (2019) 011002 doi:10.1088/1742-6596/1183/1/011002

Peer review statement

All papers published in this volume of *Journal of Physics: Conference Series* have been peer reviewed through processes administered by the proceedings Editors. Reviews were conducted by expert referees to the professional and scientific standards expected of a proceedings journal published by IOP Publishing.

IOP Conf. Series: Journal of Physics: Conf. Series **1183** (2019) 012001 doi:10.1088/1742-6596/1183/1/012001

Research of the influence of the 2D and 3D surface roughness parameters of bearing raceways on the vibration level

S Adamczak, P Zmarzły

Kielce University of Technology, al. Tysiąclecia Państwa Polskiego 7, 25-314 Kielce, Poland

E-mail: pzmarzly@tu.kielce.pl

Abstract. The level of generated vibrations is one of the most important exploitation parameters of rolling bearings. The vibration analysis allows for the evaluation of the technical condition of new and operating bearing assemblies. There are numerous factors affecting the level of generated vibrations. The condition of topography of active surfaces of rolling bearings is one of them. The state of the surface texture may be described with the use of 2D and 3D roughness parameters. Since there are no unanimous guidelines, the paper's authors decided to analyse whether, when evaluating topography of the active surface of rolling bearings it is beneficial to measure 3D roughness parameters, or whether the evaluation of 2D roughness parameters fully suffices. Fifth pieces of type 6304 ball bearings were examined. The measurement of selected 2D and 3D roughness parameters of the bearings races was performed using a contact method on a Form Talysurf PGI 1230 device made by Taylor Hobson. The measurement of vibrations generated by the bearings was performed with an Anderometer, a device developed by the employees of the Kielce University of Technology. Statistical analysis based on the correlation calculation was used to evaluate the impact of 2D and 3D roughness parameters of active surfaces of rolling bearings on the level of generated vibrations.

1. Introduction

Rolling bearings are a fundamental group of mechanical elements. Most mechanisms that rotate use rolling bearings. They are applied in almost all devices and mechanisms, from white goods, elements of machine tools, through elements of jet engines [1-3]. The functioning of entire mechanisms sometimes depends on the correct operation of ball bearings. Consequently, the technical condition of rolling bearings must be examined thoroughly to predict their wear and to schedule their replacement, if needed. In production processes, an unforeseen malfunction of bearings may result in the stoppage of an entire production line, thus generating high costs.

The level of generated vibrations is considered to be the rolling bearings' basic exploitation parameter. The vibration analysis proves to be a powerful diagnostic tool that allows to determine which component of a bearing generates the dominating level of vibrations, in consequence allowing the level of wear to be evaluated. The measurement of the level of vibrations is commonly used due to its non-destructive character. Prior to leaving the factory, each bearing is subject to vibration analysis to determine whether it can be released for use. We can also measure the vibrations of entire bearing assemblies, so that monitoring the technical condition of working bearings is possible [4]. The vibration analysis has been taken up as a subject of many scientific projects [5-7].

IOP Conf. Series: Journal of Physics: Conf. Series **1183** (2019) 012001 doi:10.1088/1742-6596/1183/1/012001

One of the main reason affecting the level of generated vibrations is the condition of the geometrical structure of the rolling bearings' active surfaces, such as the inner and outer race. Size reduction and constant improvements in the performance of devices and mechanism require that bearings support higher and higher loads, have lower and lower frictional resistance and are increasingly quieter, while maintaining long life cycle. Therefore, bearings manufacturers and scientists never stop researching ways to optimize the geometry of active surfaces of bearings in order to obtain products that satisfy the requirements of the modern technology. Consequently, the requirements relating to the modern measurement technology and methods of analyzing signals also grow [8, 9].

According to ISO 25178-6:2011 [10,11] the surface geometric structure can be examined using line-profiling method (2D roughness) and areal-topography method (3D roughness). 2D roughness parameters are commonly used in industrial and laboratory environments for the purpose of evaluating the condition of the surface texture of various products. Such parameters are measured easily and quickly, with portable measuring devices, on-site, in industrial conditions [12]. Still, the condition of the surface should be examined more thoroughly in certain cases. At such times, 3D roughness measurement and analysis of parameters are recommended [13]. Specialized, costly measuring devices is needed to conduct such measurements. It may take several hours to measure the 3D roughness of a single surface. Hence, this type of measurement is not fit for quick control of the manufactured elements of rolling bearings. Moreover, advanced knowledge and rich experience are required to interpret the results of such measurements, meaning that highly qualified personnel must be employed [14]. Additionally, in case of elements produced in additive technologies, the measurement of 3D roughness parameters is either difficult or downright impossible [15].

Currently, the most popular 2D roughness parameter, namely the arithmetical mean roughness value Ra, is used in the bearing industry to evaluate the surface quality of the raceways and the balls. Despite its big popularity, the parameter does not allow for the evaluation of the unevenness of the surface [11]. Still, there are no standards that recommend the use of other roughness parameters. The development of modern measurement techniques allows for the analysis of multiple roughness parameters, both in 2D and 3D aspect, which are related to the examined product's operating properties. This is especially important in the case of new series of bearings, where the production process may be unstable. As there are no standards or clear guidelines specifying the type of roughness parameters that should be examined by the bearings manufacturers, the paper's authors decided to analyze whether it is necessary to apply the time-consuming and costly analysis of roughness parameters of rolling bearings' surfaces in 3D, or the 2D analysis is sufficient.

2. Experimental examinations

The examinations presented in this paper were conducted in the laboratories of the Kielce University of Technology, namely in the Laboratory of Computer Measurements of Geometric Quantities and the Laboratory of Rolling-Elements Bearings. Fifty pieces of type 6304 ball bearings were examined.

2.1. Examined bearings

In order to eliminate the impact of the factory-used lubricant on the level of generated vibrations, a decision was made to use type 6304 open rolling bearings for measurements. Still, to ensure an oil film is present, a few drops of machine oil were added before the measurement of vibrations. This type of bearings is commonly used in the machinery building industry. Figure 1 presents an image of the examined bearing.

Figure 1. Example of the examined type 6304 ball bearing

2.2. Vibration analysis

The first step of the examination procedure was to measure the vibration velocity generated by the bearings. An Anderometer measuring stand, developed at the Kielce University of Technology within the framework of a project financed from the resources of the National Centre for Research and Development of Poland, was used for this purpose. The photography of the device along with one of the examined bearings is shown in figure 2.

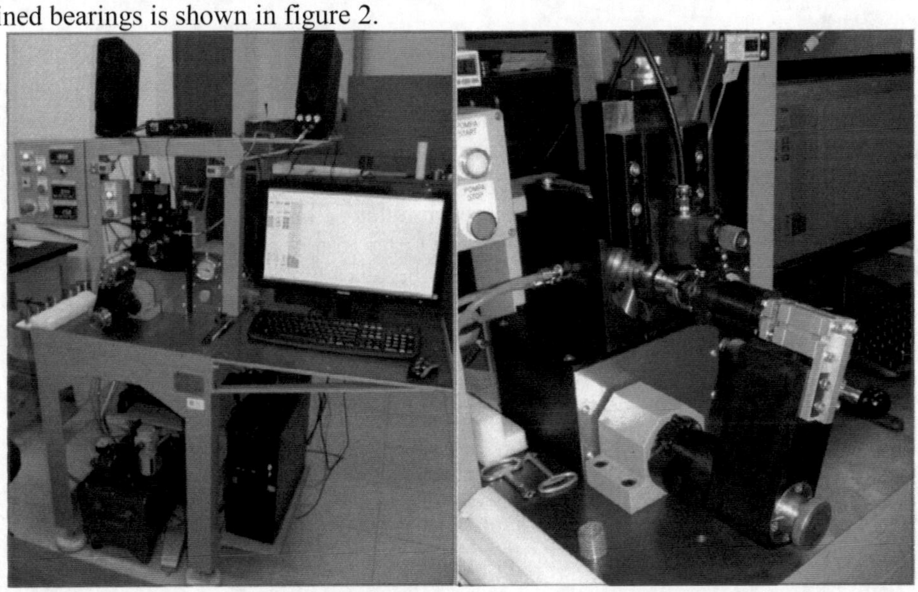

Figure 2. Anderometer, a device for measuring the vibrations of rolling-element bearing.

The examined bearings are fixed to a measuring mandrel and a pneumatic clamp ensures a constant pressure force of 65N. To eliminate the impact of external vibrations, the spindle is lubricated with an oil mist. Through a belt transmission gear, the motor rotates the spindle at 1800 min. This speed is consistent with the standard on the measurement of vibrations. The spindle's rotational speed may be freely adjusted and controlled via the digital display. A piezoelectric sensor for measuring the vibration velocity was used in the device. In comparison to the vibrations acceleration sensors commonly used in such devices, the applied sensor operating in the velocity function has a number of advantages [6, 16]. The obtained vibration signal is divided into three basic frequency bands: low (LB) from 50 to 300 Hz, medium (MB) from 300 to 1800 Hz and high (HB) from 1800 to 10 000 Hz. The level of generated vibrations is expressed with a unit called Anderon [6, 16].

IOP Conf. Series: Journal of Physics: Conf. Series **1183** (2019) 012001 doi:10.1088/1742-6596/1183/1/012001

2.3. Measurement and analysis of surface topography parameters

The measurements of roughness parameters of the raceway of type 6304 rolling bearings were taken with the Form Talysurf PGI 1230 measuring system. This is a professional device for precise evaluation of topography of the examined surface in 2D and 3D. Due to its high measurement accuracy the system is especially recommended to manufacturers of high precision rolling bearings [17]. Four basic 2D roughness parameters (Ra, Rt, Rsk, Rku) were used for the purpose of the examination, along with the relevant roughness parameters analyzed in 3D (Sa, St, Ssk, Sku). The 3D surface measurements were performed by scanning the consecutive cross-sections of the measured surface with a measuring probe. The scanning surface was 1.35 x 1.35 mm. The table 1 shows information related to conditions of measurement.

Table 1. Measurement conditions

Sampling interval	ΔX=0.05μm, ΔY=10μm
Number of repetitions	125
Cut-off value	0.25 mm
Kind of filter	Gaussian filter (ISO 11562)
Form removal method	third-order polynomial

3. Research results

The research results were presented in tables and as diagrams showing the dependence of the level of vibrations from the selected 2D and 3D roughness parameters. Table 2 shows the overall results of measurements, where $\bar{x}$ is mean value, x_{max} is a maximum value, x_{min} is a minimum values, s is a standard deviation, R is a range (difference between x_{max} and x_{min}). The results in the first main column apply to the measured level of the generated vibrations expressed in Anderon units [6]. The results of measurement of vibrations were presented for three frequency bands: low (LB), medium (MB) and high (HB). The second column presents 2D roughness parameters and the third column presents 3D roughness parameters.

Table 2. Research results

	Vibration level, Anderon			Roughness 2D				Roughness 3D			
	LB	MB	HB	Ra, μm	Rt, μm	Rku	Rsk	Sa, μm	St, μm	Sku	Ssk
$\bar{x}$	8.70	8.92	8.42	0.03	0.56	10.15	-1.57	0.03	1.25	31.10	-2.56
x_{max}	14.48	14.79	13.93	0.05	0.95	18.00	-0.29	0.04	2.81	76.90	-0.38
x_{min}	4.52	5.12	3.33	0.02	0.34	4.62	-2.85	0.02	0.69	5.89	-7.51
s	2.343	2.750	2.313	0.006	0.136	3.014	0.612	0.006	0.481	35.764	1.487
R	9.96	9.67	10.60	0.03	0.61	13.38	2.56	0.02	2.12	71.01	7.13

The selected results were presented on diagrams for the purpose of a broader analysis and to determine the dependences between the 2D and 3D roughness parameters and the amplitudes of the generated vibrations. Therefore, the parameters Ra and Sa are widely used in laboratory and industrial application the Figure 3 presents the relationship between the level of generated vibrations and the Ra parameter of the race's roughness profile while Figure 5 shows impact of Sa parameter on the vibration level.

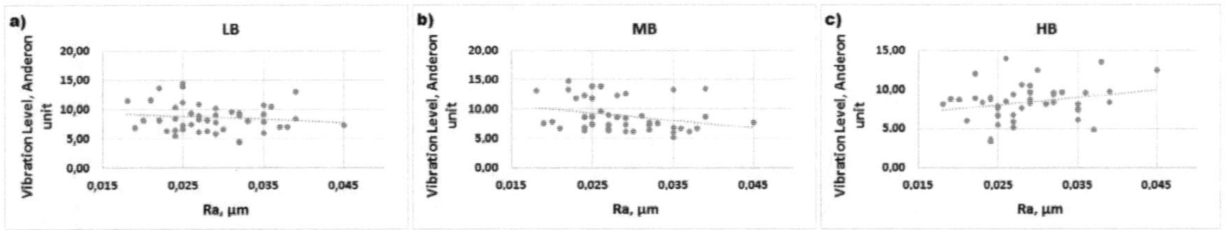

Figure 3. Diagrams of the relationship between the level of generated vibrations and the arithmetical mean roughness value (Ra) of the race's roughness profile, where: a) LB - low frequency bandwidth, b) MB - medium frequency bandwidth, c) HB - high frequency bandwidth.

The analysis of results presented on diagrams in figure 3 allows for the drawing of a conclusion that the Ra roughness parameter ranging from 0.02÷0.05 μm was obtained for the races of the examined bearings. The standard deviation was s=0.006 μm. This indicates low variability of the Ra parameter for the series of the examined bearings. Examining the impact of the arithmetical mean roughness value on the level of generated vibrations, a conclusion can be made that an increase in the Ra parameter causes a slight drop of the level of generated vibrations in the low and medium frequency band (see figure 3a and 3b). However, as the Ra parameter increases, the level of generated vibrations registered in the high frequency band shows an increasing tendency (figure 3c).

Additionally Figure 4 shows example of surface roughness profile characterized by various values of the Ra parameters.

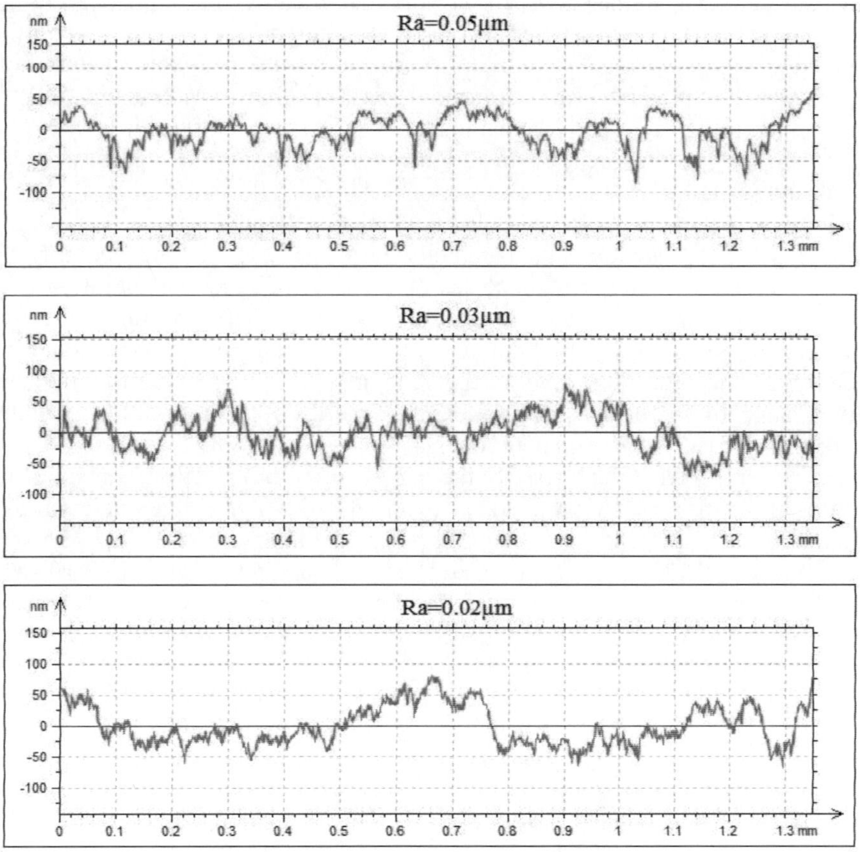

Figure 4. Example of surface roughness profile.

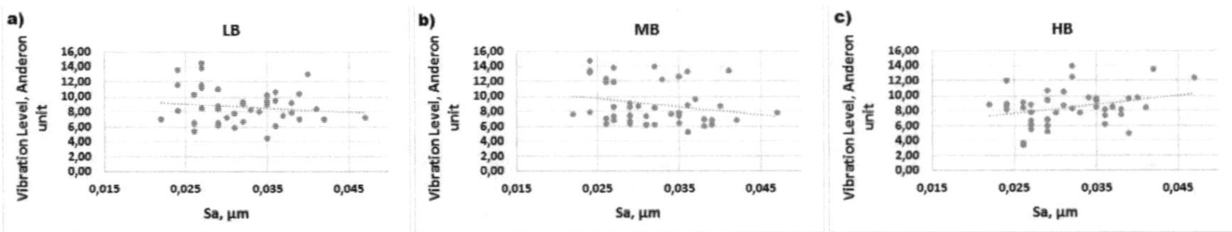

Figure 5. Diagrams of the relationship between the level of generated vibrations and the arithmetical mean height (Sa) of the race's surface, where: a) LB - low frequency bandwidth, b) MB - medium frequency bandwidth, c) HB - high frequency bandwidth.

Analyzing the stereometry of the surface of the bearings' races it may be concluded that the range of the obtained Sa parameters was 0.02÷0.05 µm. In the case of analysis of the 3D roughness parameter i.e. Sa it may be concluded that the same value and the mean square error were calculated for the Ra parameter (see table 2). In this case, too, an increase in the roughness parameter Sa causes a slight drop of the vibrations measured in low (LB) and medium (MB) ranges of frequency. An increasing tendency, however, was noted for the high frequency band.

When comparing the values of the measured roughness parameters Ra with their counterparts in the aspect of 3D measurements, namely the Sa parameters, a conclusion can be made that the results overlap to a large extent. In relation to this, when analyzing the Ra and Sa parameters, there is no need to measure the 3D roughness parameters because the 2D parameter yields similar results.

Another group of the analyzed parameters were the total height of the roughness profile Rt and the total height surface roughness St. The variability range of the parameter determining the total roughness height was within range Rt∈<0.34µm; 0.95µm>. Analyzing research results it may be concluded that an increasing the roughness parameter Rt causes an increase in the level of generated vibrations across all analyzed frequency bands. However, the largest impact was registered for the mean frequency band - MB.

As per the European standard [18], the equivalent of the Rt parameter in 3D is the total height surface roughness St. Consequently, this parameter was also used to analyze the impact of the parameters of the roughness of the bearing raceway on the level of generated vibrations.

When analyzing the measured values of the St parameter that determines the height surface roughness, a conclusion can be made that, similarly to the Rt parameter, an increasing the value of the St parameter causes an increase in the level of the generated vibrations in the low frequency band LB and the medium range of frequencies. However, a slight drop of the amplitude of the generated vibrations can be observed as the St parameter rises.

While comparing parameters which specify the height of the profile of roughness of the examined races of rolling bearings, namely Rt parameters measured in 2D and St parameters measured in 3D, one can conclude that the results show a higher divergence than in the case of Ra and Sa parameters. A mean square error of s=0.136 µm was obtained for Rt parameters, and of s=0.481 µm for St parameters. This proves that the dispersion of the determined parameters is significantly broader. Moreover, this dispersion was broader for the St parameter which determines the total height of surface roughness.

Functional parameters were analyzed in order to describe in detail the nature of surface texture of examined bearings. The first group comprised parameters describing the flattening of roughness peaks (kurtosis) i.e. Rku and Sku. These parameters are especially important when evaluating the tribological properties of rolling bearings, because they can be related to friction coefficient [19].

Analyzing research results a conclusion can be made that the values of the Rku parameters show a high divergence. This is confirmed by the value of the determined standard devotion of s=3.014 at an average value of $\bar{x} = 10.15$ (see table 2). The value of kurtosis of Rku>3 was obtained for all

IOP Conf. Series: Journal of Physics: Conf. Series **1183** (2019) 012001 doi:10.1088/1742-6596/1183/1/012001

measured races, proving the presence of sharp peaks and grooves in the roughness profile. This is confirmed by the exemplary roughness profiles that were shown in Figure 4.

Similarly to Rku parameters, Sku parameters show a high divergence. The variability range of the distribution concentration ratio of the height of surface topography was within the range Sku = 15÷175.

The summary of analysis of the impact of Rku and Sku roughness parameters on the level of generated vibrations fails to yield unanimous conclusions. The examined parameters determining the character of roughness peaks exhibited very high divergence. The standard deviations achieved high values both for the parameters derived from the 2D profile, i.e. Rku, and for the 3D parameters, i.e. Sku.

The Rsk parameter is an important parameter determining the skewness of the roughness profile. This parameter can be related to friction coefficient. Consequently, it is advisable to use such a parameter to evaluate the topography of the surface of rolling bearings in terms of friction and wear. The equivalent of the Rsk parameter in 3D is Ssk (skewness).

A conclusion can be drawn from the analysis of results table 2 that negative values of Rsk parameters were obtained for all examined bearing races. This means that material is concentrated in the vicinity of peaks of the roughness profile, which indicates surfaces whose shape is similar to plateaus. The negative value of the asymmetry factor of the roughness profile also indicates the presence of "deep valleys" in the surface of the examined element. Such a nature of surface allows for the better retention of a lubricant, indicating good bearing surface. The analysis measuring results leads to a conclusion that as the value of the Rsk parameter increases, the value of generated vibrations recorded in the low and medium frequency range remains at a constant level. However, when considering the high range of vibration frequency, a slight drop in the amplitude of the generated vibrations was observed, along with an increase in skewness of the roughness profile.

The parameters describing the profile skewness are difficult to link to the values of vibrations generated by the examined bearings, because despite the changes of the values of the analyzed parameters, the values of the generated vibrations would remain at a constant level. Still, the Rsk parameter may serve to evaluate the shape of the peaks of the roughness profile, which is of crucial importance when evaluating the tribological properties of the surface of bearings. Therefore, it seems justified to analyze the values of parameter Rsk when evaluating the frictional moment of rolling bearings.

The equivalent of the Rsk parameter for the purpose of analysis of surface topography is the Ssk parameter. This parameter may be used to evaluate the friction and wear of the active surface of rolling bearings. Similarly as for the Rsk parameter, the values of topography skewness (Ssk) of rolling bearings are negative. The Ssk parameter is interpreted similarly to the Rsk parameter, meaning that negative values indicate a surface with plateau-shaped roughness peaks.

However, in the case of the equivalent of the parameter in 3D we observe a relationship between the values of Ssk parameters and the values of generated vibrations. An increase in the value of the Ssk parameter causes a decrease in the value of generated vibrations across all analyzed frequency ranges. The impact is most evident for the medium frequency band.

A conclusion can be drawn on the basis of the analysis of parameters of skewness of profiles and roughness topography that determining the 3D parameter, namely Ssk, seems justifiable. This parameter allows for the description of the nature of the peaks of unevenness with higher complexity, thus making it easier to evaluate tribological properties.

Pearson correlation coefficients (see table 3) were calculated in order to quantitatively evaluate the relationship between the determined 2D and 3D roughness parameters and the level of vibrations generated by the rolling bearings. Additionally, the correlation between the 2D roughness parameters and their equivalents analyzed in 3D aspect was examined. The results were presented in Table 4.

Table 3. Matrix of Pearson correlation coefficients between respective roughness parameters and the level of generated vibrations

	Ra	Sa	Rt	St	Rku	Sku	Rsk	Ssk
LB	0.233	0.205	0.275	0.087	0.133	0.133	-0.116	-0.188
MB	-0.218	-0.214	-0.065	0.061	0.015	0.102	0.192	-0.016
HB	0.112	0.123	0.021	-0.066	-0.154	-0.018	0.075	0.041

The analysis of the calculated values of Pearson linear correlation coefficients presented in Table 3 permit a conclusion that weak or low correlations could be observed between the 2D and 3D roughness parameters and the level of measured vibrations of rolling bearings. Considering the Ra and Sa parameters, higher correlation between the measured 2D parameters can be seen than between the 3D parameters. The high frequency band is an exception. This confirms the previous finding that there is no need to conduct time-consuming measurements and analysis of the Sa parameter when analyzing the raceways of rolling bearings. Similar results were obtained for parameters specifying the total height of the roughness profile Rt and St. Higher correlative dependence was obtained for the Rt parameter than for the St parameter. When considering the Rku and Sku parameters describing the skewness of the roughness profile, it can be observed that Pearson correlation coefficients assumed the same value for the low range of vibration frequency (LB). However, a higher correlation for the parameter Sku was obtained for the medium range of vibration frequency (MB). The analysis of the high range of vibration frequency (HB) shows higher correlative dependence for the 2D parameter, Rku. The interpretation of results for parameters describing the skewness of the roughness profile, namely Rsk and Ssk, is inconclusive. In the case of low band of vibration frequency, higher correlation was achieved for Ssk parameters. However, higher correlation is shown by 2D parameters, namely Rsk, for vibrations analyzed in the medium and high frequency range.

Table 4. Matrix of Pearson correlation coefficients between individual 2D and 3D roughness parameters.

	Ra
Sa	0.96

	Rt
St	0.50

	Rku
Sku	0.27

	Rsk
Ssk	-0.04

Considering the results of calculations presented in table 4, a conclusion may be drawn that the highest correlation exists between the group of amplitude parameters, namely Ra and Sa parameters. A very certain correlative relationship was achieved in this case. A significant correlative relationship was observed among parameters specifying the total height of the roughness profile, i.e. Rt and St. Worse results were obtained for Rku and Sku parameters, where low correlation was found. The correlative relationship between 2D and 3D parameters was almost insignificant for parameters determining the profile's skewness (Rsk and Ssk).

It should be noted that the discrepancies between parameters Rt, Rku, Rsk and their 3D extension may result from the number of measurement points. It is know, that the number of measurement points used areal-topography method is greater than in line-profiling method.

4. Summary

The dynamic development of the bearing industry results in the growth of qualitative requirements of the manufactured products with the simultaneous reduction of production costs. To satisfy the requirements of the modern market the manufacturers must improve their measurement procedures, apart from constantly increasing the efficiency of production. This is of special importance in the case of analysis of the topography of active surfaces of bearings, because its quality affects the operating properties of bearing assemblies. There is no one universal parameter that can sufficiently characterize the condition of the surface texture of the elements of rolling bearings. Currently, the surface roughness can be analyzed with line-profiling method (2D roughness) or areal-topography method (3D roughness). The selection of a suitable measurement method and the selection of roughness parameters that allow for the sufficient evaluation of the condition of the surface layer of rolling bearings is not fully unequivocal. It is recommended to select such roughness parameters that can be correlated with other operating parameters of rolling bearings, e.g. the level of generated vibrations. There are no generally available standards and guidelines determining the type of roughness parameters to analyze.

Thanks to the development of measurement systems which allow for the comprehensive analysis of the surface topography, 3D roughness measurements are gaining popularity. However, measuring a single surface using contact methods may take up to several hours. Consequently, the paper presents examinations whose purpose is to specify which 2D and 3D roughness parameters can be related to the level of generated vibrations.

Based on the research results presented in this paper, a conclusion can be made that the time-consuming analysis of surface topography of the raceways of rolling bearings is not always justified. Considering the amplitude roughness parameters, namely Ra and Sa, most commonly used in industrial conditions, it can be observed that it is the 2D roughness parameters (Ra) that exhibit higher correlation with the level of vibrations generated by the bearing. Similar results were obtained for the Rt and St parameter. Consequently, 2D profile analysis is sufficient for such parameters. However, when considering the results of examinations obtained for parameters describing in detail the surface texture, such as Rsk, Ssk, Rku and Sku, it is recommended to analyze the roughness parameters determined on the basis of 3D surfaces.

Summing up the results of the examination a conclusion may be drawn that the analysis of 2D roughness parameters, namely Ra and Rt, is sufficient for the more stabilized production processes. However, when beginning the production of new types of bearings or when modifying the processing parameters, it is recommended to analyze the 3D roughness parameters which describe the nature of surface topography in finer details.

The examinations presented in this paper are of initial research. The direction of further examinations will consist in expanding them with further 2D and 3D roughness parameters. Additionally, other types of rolling bearings will be examined. Research results indicate low and weak correlation between roughness parameters and vibration level. Therefore, the impact of the other factors (roundness and waviness deviation, internal clearance or lubrication types) on the vibration level should be investigated

Acknowledgments

The study was possible thanks to a project contract with Polish National Centre of Research and Development no. PBS2/A6/20/2013/NCBiR/24/10 /2013: “Research and evaluation of reliability of modern methods of surface topography measurements in micro and nano scale”.

References

[1] Gupta P and Pradhan M K 2017 *Materials Today: Proceedings* **4** 2085–2094.
[2] Nowakowski L, Miko E and Skrzyniarz M 2016 *Engineer Mechan* **2016** 426-429.
[3] Wang H, Han Q, Luo R and Qinq T 2017 *J Sound Vib* **406** 124–145.
[4] Stanik Z 2014 *Solid State Phenomena* **210** 58-64.
[5] Randall R B and Antoni J 2011 *Mech Syst Signal Pr* **25** 485–520.
[6] Adamczak S and Zmarzły P 2017 *Bull Pol Acad Sci-Te* **65(4)** 541-551.
[7] Rai A and Upadhyay S H 2016 *Tribol Int* **96** 289–306.
[8] Adamczak S, Zmarzły P and Stępień K 2016 *Bull Pol Acad Sci-Te* **64 (2)** 45-52.
[9] Makieła W and Gogolewski D 2017 *Procedia Engineer* **192** 243-248.
[10] ISO 25178-6: Geometrical product specifications (GPS) - Surface texture: Areal - Part 6: Classification of methods for measuring surface textur.

[11] Mathia T G, Pawlus P and Wieczorowski M 2011 *Wear* **271** 494–508.
[12] Stępień K 2015 *Key Eng Mater* **637** 69-73.
[13] Hoła J, Sadowski Ł, Reiner J and Stach S 2015 *Constr Build Mater* **84** 111–120.
[14] Miller T, Adamczak S, Świderski J, Wieczoroski M, Łętocha A and Gapiński B 2017 *Bull Pol Acad Sci-Te* **65 (1)** 53-61.
[15] Kozior T Kundera Cz 2017 *Procedia Engineer* **192** 463-468.
[16] Adamczak S, Domagalski R, Sender E, Zmarzły P and Gorycki Ł 2016 *Diagnostyka* **17(1)** 41-49.
[17] http://pdf.directindustry.com/pdf/taylor-hobson/form-talysurf-pgi-830-1230/7159-390621.html
[18] ISO 25178-2: Geometric Product Specifications (GPS) – Surface texture: areal.
[19] Grzesik W 2016 *J Mater Eng Perform* **25(10)** 4460-4468.

Evaluation of the influence of selected parameters of Selective Laser Sintering technology on surface topography

Cz Kundera[1], T Kozior[1]
[1]Kielce University of Technology, Faculty of Mechatronics and Mechanical Engineering, Department of Manufacturing Engineering and Metrology, Poland

E-mail: tkozior@tu.kielce.pl

Keywords: GPS, Additive Technologies, SLS, PA 2200

Abstract. The paper presents the preliminary research results of studies of the influence of basic technological parameters of the additive technology, Selective Laser Sintering (SLS), on the quality of geometric surface texture. Microstructural studies included the determination of two-dimensional as well as three-dimensional parameters of surface texture. The study includes the influence of technological parameters such as the orientation of the models on the virtual building platform, the energy density transmitted to the sintered powder layer and the building layer thickness. The results confirm that the above mentioned parameters have a significant influence on the geometrical surface texture. The most favorable variants of the location of the models on the building platform, the value of layer thickness and energy density applied to sintered powder were determined.

1. Introduction

Additive manufacturing (AM) technologies have revolutionized the manufacturing industry over the last few years. Because of the layered nature of the building process and possibility to manufacture models directly from 3 dimensional CAD files, AM allows production of unique models with almost unlimited shapes, especially with complex internal dimensions. Generative technologies are successfully found in many kinds of industries as both prototype and fully functional applications. The ever-expanding branch of both the production of more accurate machines and modern construction materials used in the layered technologies shows that these technologies are becoming more competitive in some applications than conventional manufacturing methods [1] such as machining, forging or casting.

Selective Laser Sintering of polyamide powders is one of the most important additive manufacturing technologies. Due to the use of material in a form of powder which is based on construction polyamide (PA), SLS technology has great application potential [2]. This applies both to the production of prototype models [3,4] and fully functional components of machines and equipment, for example technical seals or bearing arrangements [5,6,7]. In this technology there is no need to build support structures, which significantly increases its production potential. The recent trend in the development of additive technologies demonstrates that these technologies will continue to evolve dynamically in the direction of hybrid machines (combination of CNC machining centres and additive machines) and with the development of the so called "overprinting" to obtain composite elements [8].

With the development of technology and consequently the complexity of construction of machines implementing layered technologies, a significant problem has arisen. There is an increase in the number of process parameters which influence the properties of manufactured elements. These include, especially, dimensional accuracy and mechanical properties, which directly influence performance characteristics such as abrasive wear processes [9]. Therefore, it is crucial to determine the impact of process parameters on the above mentioned properties. This is particularly true of the surface texture quality due to its strong influence on the correct co-operation of the associated machine elements. Experimental research is focused on both mechanical properties [10,11,12] and accuracy as well as mathematical analysis of the surface texture [13,14,15,16,17]. These papers describe the influence of process parameters on mentioned phenomena.

The paper [14] presents the research results, which consisted of determining the influence of selected process parameters on the microstructure of the examined surface. In addition further laser surface treatment for samples made of metal powder manufactured by SLS technology were included. Research has shown that it is possible to improve (reduce) the value of the surface texture parameters by laser treatment, and the key technological parameters are laser power and energy density.

Preliminary studies conducted by the authors of the paper [9] have shown that the "printing" direction on the platform during manufacturing and the building layer thickness affect the abrasive wear process. The research was carried out on the well-known polyamide PA 2200 using a tribometer (ring on disc type).

Research conducted in the work [16] was based on the influence of selected process parameters such as laser power, scan spacing, bed temperature and hatch length on 2D geometric surface parameters: *Ra*, *Rz* and *Rq*. The material used for the construction of the samples was Dura-form polyamide. Research has shown significant impact of the aforementioned technological parameters on the measured quantities, in particular the laser power, where, with increasing its value, the roughness parameters decrease.

The important scientific topic of the influence of technological parameters of selective laser sintering technology on the dimensional accuracy of the models, especially technological surface texture is discussed in the presented paper. Further research will involve the extension of research especially the impact of process parameters on the more complex 3D surface texture parameters, the mass of building models and stress relaxation during compression testing.

2. Research and Methodology

Cylindrical samples with a diameter of 13 mm and 6.3 mm high were designed in SolidWorks, CAD software. These dimensions are in accordance with the ISO standard [18] for specimens designed to determine stress relaxation during uni-axial compression tests. By using EOS Formiga P100 machine, which is a selective laser sintering technology printer, physical models of the sample with predefined technological parameters have been manufactured. Each type of sample was made in 3 pieces. Three characteristic variants of positioning of models on the machine building platform, i.e. at 0 °, 45 ° and 90 °, were selected. The building layer thickness was equal to 0.1 and 0.2 mm. All type of sample was produced with three specified values of energy density i.e. 0.032 J/mm^2, 0.056 J/mm^2 and 0.08 J/mm^2. Material used to construct the samples was polyamide PA 2200 which is based on the well-known construction material polyamide PA 12. Measurements of the surface texture were carried out at the Laboratory of Computer Measurement of Geometric Quantities at the Kielce University of Technology using the Taylor Hobson Talysurf CCI Lite. This devise uses coherence correlation interferometry measurement technique (CCI). This precision optical measurement profilometer is characterized by a measuring range of 2.2 mm, with a vertical resolution of Z – 10 pm and optical resolution (X, Y) of 0.4-0.6 μm [19]. The parameters of primary profile (*Pa*, *Pz*) and surface (*Sa*, *Sz*) were obtained after leveled processes. To compare research results further filtering operation was not necessary. Surface geometry studies were carried out at the top surface of the sample (Fig.1) using the magnification equal to 20, where the measured area was 0.8 x 0.8 mm. The geometric parameters [20] of the primary 2D profile (*Pa* and *Pz*) and 3D geometric parameters (*Sa* and *Sz*) were determined for the comparison of the research results (with different technological parameters).

IOP Conf. Series: Journal of Physics: Conf. Series **1183** (2019) 012002 doi:10.1088/1742-6596/1183/1/012002

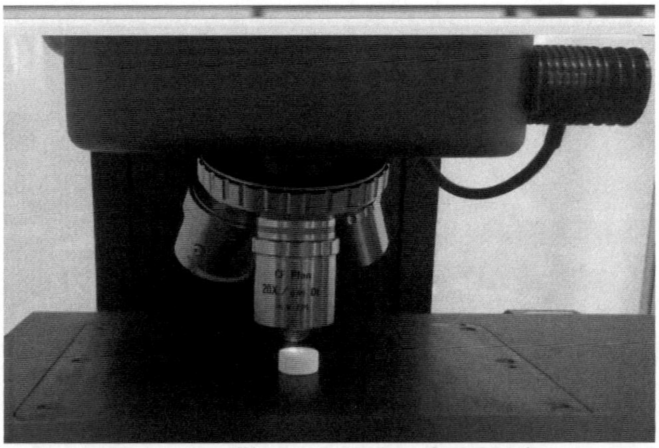

Fig. 1. Profilometer, Talysufr CCI Lite

3. Results

The results of the studies presented in the table 1 show the mean values of the three measured samples for the each type. The table includes both 2D and 3D surface texture parameters and manufacturing technological parameters where: *LT* - layer thickness, *PD* – printing direction, *ED* – energy density.

Table 1. Texture parameters

No.	*LT*, mm	*PD*, °	*ED*, J/mm²	*Sa*, µm	*Sz*, µm	*Pa*, µm	*Pz*, µm
1	0.1	0	0.032	13.534	108.347	13.175	74.976
2	0.1	0	0.056	13.692	94.598	13.359	77.8
3	0.1	0	0.08	13.637	115.013	13.04	74.692
4	0.1	45	0.032	14.276	116.467	13.44	76.695
5	0.1	45	0.056	17.258	139.527	16.404	92.979
6	0.1	45	0.08	14.088	106.948	13.444	75.078
7	0.1	90	0.032	12.925	101.493	12.584	72.251
8	0.1	90	0.056	14.121	110.733	13.285	77.487
9	0.1	90	0.08	14.554	117.37	13.981	79.12
10	0.2	0	0.032	12.964	108.383	12.574	73.444
11	0.2	0	0.056	13.179	107.707	12.765	75.886
12	0.2	0	0.08	12.019	97.205	11.619	68.076
13	0.2	45	0.032	19.937	126.673	19.421	96.458
14	0.2	45	0.056	13.67	106.983	13.113	76.005
15	0.2	45	0.08	14.445	121.017	13.532	77.689
16	0.2	90	0.032	18.706	122.097	18.017	91.491
17	0.2	90	0.056	13.708	106.747	13.01	74.286
18	0.2	90	0.08	16.149	123.037	15.328	84.884

IOP Conf. Series: Journal of Physics: Conf. Series **1183** (2019) 012002 doi:10.1088/1742-6596/1183/1/012002

a)

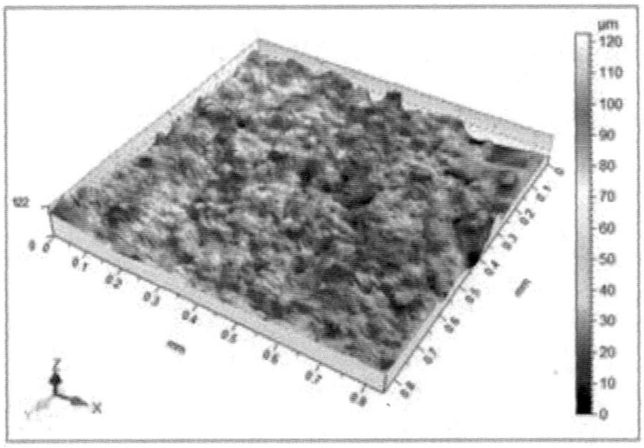

b)

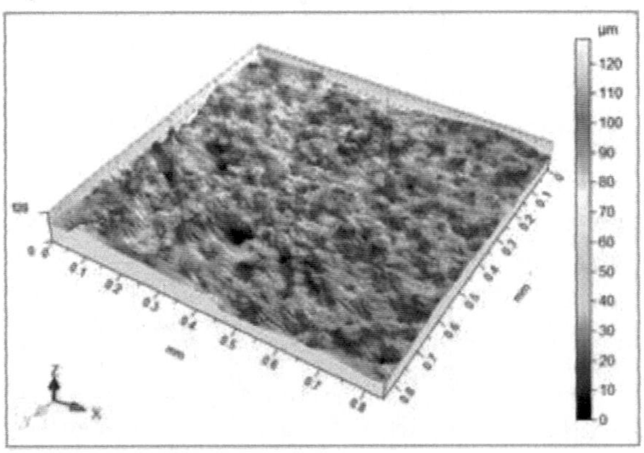

c)

Fig. 2. 3D view of tested surfaces, samples 1-9: a) 0 °, b) 45 °, c) 90 °

IOP Conf. Series: Journal of Physics: Conf. Series **1183** (2019) 012002 doi:10.1088/1742-6596/1183/1/012002

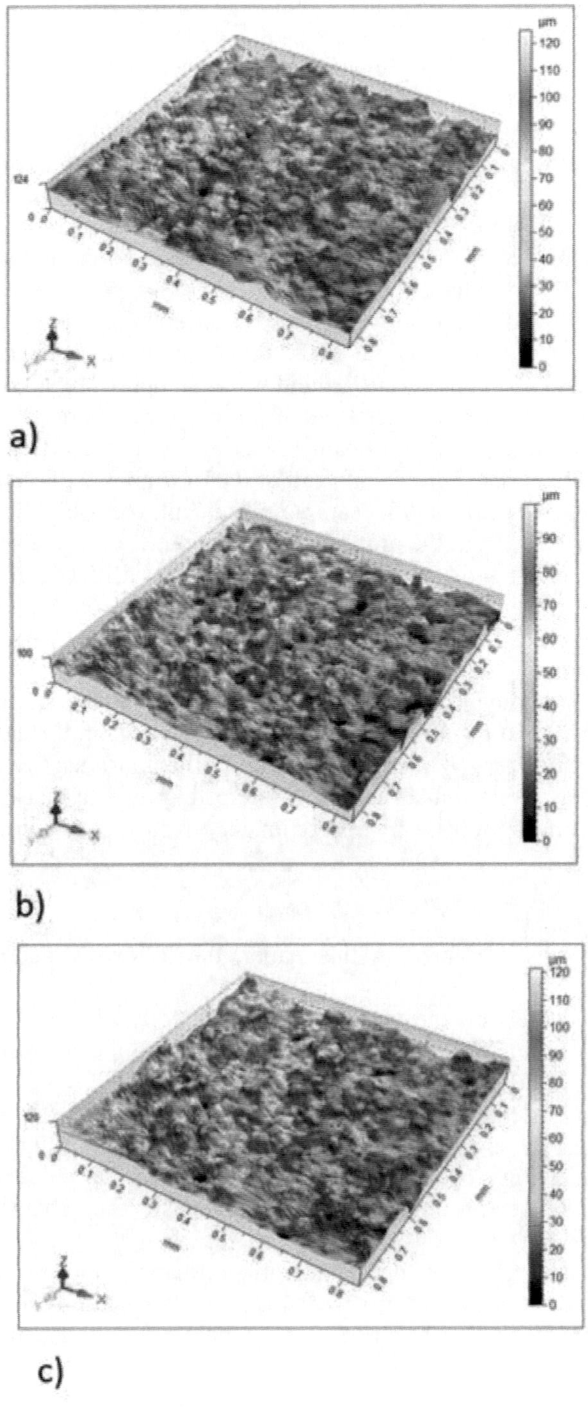

Fig. 3. 3D view of tested surfaces, samples 10-18: a) 0 °, b) 45 °, c) 90 °

IOP Conf. Series: Journal of Physics: Conf. Series **1183** (2019) 012002 doi:10.1088/1742-6596/1183/1/012002

Quantitative analysis of the research results, shows that the technological parameters significantly affect the quality of the surface geometry. There is a noticeable effect on the thickness of the building layer and the position on the platform "printing direction" as well as the density of energy transmitted to the sintered layer.

Surface texture views made using the optical profilometer shown in Figures 2 and 3 indicate a phenomenon of the so called stairs-step effect. This phenomenon is noticeable in Figures 2b and 3b, where samples were manufactured with a positioned at an angle of 45 °. This indicates a large area of valleys and peaks at the contact points of the layers being built.

Samples from 1 to 9 manufactured with a layer thickness of 0.1 mm and with the lowest value of energy density equal to 0.032 J/mm^2 have the lowest values of GPS parameters. In this variant of layer thickness, one can notice the significant influence of the orientation of the models on the building platform. Samples manufactured with an angle of 0 ° do not show significant influence of the energy density on the surface texture, differences are around 1 %. Samples prepared in the direction on the building platform equal to 45 ° are influenced much more by the energy density. When comparing sample numbers 4 and 5 you can see a significant reduction in 2D and 3D parameters (about 17 %) when using an energy density of 0.032 J/mm^2 instead of 0.056 J/mm^2 (0.056 is recommended value by the machine producer). In the case of the 90 ° sample placement, the use of a similarly lower value of energy density results in a reduction in the aforementioned parameters values of approximately Sa and Sz around 8 %, Pa – 5 %, and Pz – 6 %. Taking into account the influence of the "printing" direction with the same level of *ED*, it can be concluded that in almost all cases the GPS parameters have the highest values for the angle of 45 °.

Analyzing the second type of samples, i.e. those manufactured with a specified layer thickness of 0.2 mm, we can observe similar changes in the GPS parameters. There is no impact of energy density on samples with an angle of 0 °. It can be said that the increase in *ED* in the case of sample number 12 has reduced the value of parameters by about 10 %. As in other variants, here also the positioning of the samples at 45° is the least preferred variant. It is interesting to note that in the case of decreasing the value of *ED*, the GPS values increase by approximately 30% and 40% for sample numbers of 16 and 13 respectively.

4. Conclusion

Analyzing the research results presented in this paper, the following general conclusions can be formulated.

The orientation of the models on the virtual platform directly affects the quality of the surface texture. The location of the examined plane at an angle of 45 ° to the building platform results in an unfavorable increase in the value of geometric surface parameters.

The impact of the energy density parameter is related to the thickness of the layer being built. When the thickness of the layer is equal to 0.1 mm, the lower value of the energy density advantageously affects the quality of the surface texture. Samples prepared with a specified layer thickness of 0.2 mm and reduced values of the energy density are characterized by unfavorable increases in GPS parameters.

The increased thickness of the layer (up to 0.2 mm) influenced the manufacturing time, but the average value of parameters increased by up to 5% (Sa) in this case. Particularly noticeable is the increases in GPS parameters for the location at a given angle of 90 ° and layer thickness of 0.2 mm.

Acknowledgements

The study was conducted using research facilities purchased with EU funds in the framework of the 2007-2013 Development of Eastern Poland Operational Programme, LABIN Project – Support for Innovative Research Facilities of the Kielce University of Technology in Kielce. Priority 1 – Innovative Economy, Measure 1.3 – Support for R&D Projects. Therefore, the authors would like to thank to leaders of LABIN Project.

IOP Conf. Series: Journal of Physics: Conf. Series **1183** (2019) 012002 doi:10.1088/1742-6596/1183/1/012002

References

[1] Nowakowski Ł and Miko E 2012 *XIIITH Int. Conf. HERVICON-2011(Sumy)* vol **39** 395 – 404
[2] Leu M C and Guo N 2013 *Appl Mech Mater* **8** 215-243
[3] Chen Y, Zhezheng C 2011 *Rapid Prototyping J* **17** 408-417
[4] M M Musztyfaga-Staszuk 2017 *Int J Thermophys* **38**
[5] Adamczak S and Zmarzły P 2017 *Bull Pol Acad Sci-Te* **65** 541–551
[6] Kundera Cz and Bochnia J 2014 *Rapid Prototyping J* **20** 533-540
[7] Kundera Cz and Kozior T 2014 *ADV MAT RES* **874** 77-81
[8] Melnikova R, Ehrmann A, Finsterbusch K 2014 *Adv Mater Res-Switz* **62** 1-6
[9] Kundera Cz and Kozior T 2016 *Tribologia* **5** 73-84
[10] Salmoria G V, Leite J L, Veira L F, Pires A T N, Roesler C R M 2012 *Polym Test* **31** 411 – 416
[11] Salmoria G V, Leite J L, Paggi R A, Arhens C H, Pouzada A S 2012 *Int J Adv Manuf Tech* **59** 583 – 591
[12] Pilipović A et al 2010 *Int Conf iCAT2010,* (Nova Gorica, Slovenia)
[13] Kim W J, Nguyen H H, Kim H Y, Nguyen M T, Park H S and Kim J C 2017 *Arch Metall Mater* **62** 1215–1218
[14] Lamikiz A, Sanchez J A, de Lacalle L N L, Arana J L 2017 *Int J Mach Tool Manu* **47** 2040 – 2050
[15] Makieła W and Gogolewski D 2017 *Procedia Engineer* **192** 243 – 248
[16] Sachdeva A, Singh S, Sharma V S 2013 *Int J Adv Manuf Tech* **64** 1505–1516
[17] Schmid M et al 2017 *Appl Sci-Basel* **7**
[18] ISO 3385:2002 2002 Rubber, vulcanized or thermoplastic — Determination of stress relaxation in compression at ambient and at elevated temperatures.
[19] www.taylor-hobson.pl available 26.02.2018.
[20] Mathia T G, Pawlus P, Wieczorowski M 2011 *Wear* **271**, 494-508.

Calculation of plasticity index of honed cylinder liner textures

P Pawlus[1], **W A Grabon**[1] **and D Czach**[1]
Rzeszów University of Technology, Poland

E-mail: ppawlus@prz.edu.pl

Abstract. The fundamental objective of this work is to develop method of plasticity index calculation for honed and especially plateau honed cylinder textures. Cylinder liner surface topographies were measured by white light interferometer Talysurf CCI Lite. Honing angle was in the range: 15-70^0. For each identified summit the ratio of the smaller and the larger radii in the axial and circumferential directions γ_1, was computed. Then the ratios of the radii in orthogonal directions for different orientations (9^0 resolution) were computed and the smallest value, γ_2 was recorded. Average values of these ratios were compared and errors of plasticity index estimation were computed. It was found that for both honed and plateau honed liners only a tolerable level of error is introduced in plasticity index calculations by taking the cylinder axis and circumference as the principal axes. These errors were larger when the honing angles are bigger.

Keywords: surface topography, cylinder liners, plateau honing, plasticity index

Nomenclature

E'	-	Hertz elastic modulus
H	-	Hardness of the softer material
PP	-	Plateau part
VP	-	Valley part
R	-	Mean radius of summits curvature
R_e	-	Equivalent radius of curvature
S	-	Standard deviation
Spq	-	Standard deviation of plateau height
Sq	-	Root-mean-square value of the surface deviations
A	-	Honing angle
Γ	-	Degree of anisotropy
γ_1	-	Ratio of the smaller and the larger radii in axial and circumferential directions
γ_2	-	The smallest value from the ratios of the radii in orthogonal directions for different surface orientations
σ_s	-	Standard deviation of asperity height
Δ	-	Error in the plasticity index estimation using γ_1 than γ_2
$\varphi(\gamma)$	-	Correction factor
Ψ	-	Plasticity index

IOP Conf. Series: Journal of Physics: Conf. Series **1183** (2019) 012003 doi:10.1088/1742-6596/1183/1/012003

1. Introduction

Wear regimes are a function of a plasticity index, first introduced by Greenwood and Williamson [1], which determines the mode of deformation of an array of asperities of varying heights. The plasticity index has the form:

$$\Psi = \frac{E'}{H}\left(\frac{\sigma_S}{R}\right)^{1/2} \tag{1}$$

where: H – hardness of the softer material, E' – Hertz elastic modulus, σ_s – standard deviation of asperity height, R – mean radius of summits curvature.

According to Greenwood and Williamson the contact should be plastic for plasticity index larger than 1.0 and elastic for smaller than 0.6, but for the plasticity index between 0.6 and 1 the mode of deformation is in doubt. The mode of surface deformation is important in understanding contact mechanics of rough surfaces. A boundary from safe to unsafe sliding depends on the plasticity index [2, 3].

The initial version of plasticity index can be calculated on the basis of surface profiles. However, the importance of areal characterization of surface topography is being increasingly recognized. In this model, all asperity summits have the same radius but their height varied randomly. For isotropic surfaces of Gaussian height distribution, parameters characterizing surface topography can be predicted based on profile analysis using Nayak's predictions [4, 5]. Anyway, they can be calculated directly on the basis of analysis of areal surface topography [6]. However, most real surfaces are anisotropic, so there were several attempts to develop a modern version of the plasticity index, for instance for strongly anisotropic (one-directional) random surfaces [7]. The contact deformation is strongly dependent on the ratio of the principal curvature radii of asperities (degree of anisotropy) γ. Thus a general expression was developed [8] for plasticity index, taking the degree of anisotropy into consideration by assuming that all asperities are ellipsoidal near their summits:

$$\Omega = \varphi(\gamma)\frac{E'}{H}\left(\frac{\sigma_S}{R_e}\right)^{0.5} \tag{2}$$

where R_e is an equivalent radius of curvature, the geometric mean of the principal radii, and $\varphi(\gamma)$ is a correction factor of near unity if $\gamma > 0.2$.

Typically it is assumed that measurements can be taken along and across the lay (main surface direction). However, for honed cylinder liners this is not possible since they have cross-hatched structure with a two-directional lay.

There then arises a need to estimate the degree of anisotropy for cross-patterned cylinder surfaces. The other problem is that cylinder liner surface after plateau honing is a kind of two-process surfaces [9, 10, 11, 12, 13, 14] consisting of two parts: plateau and valley. The plateau honed cylinder liner surface texture is a composition of relatively smooth plateau areas ensuring low wear intersected by valleys with great ability to maintain lubricant, improving tribological performance of a piston ring pack. The quantitative assessment of two-process structures is contained in ISO 13565-3 standard, basing on the cumulative distribution plot on normal probability graph. Greenwood and Williamson [1] thought that non-Gaussian worn surfaces can be characterized by a Gaussian distribution of summit height. Leefe [15] also believed that because contact was related to the behavior of plateau surface portion, the contact behavior is governed by the Gaussian distribution of summit heights from this surface part. However it was found that the valley part affected contact properties of the stratified surface. The method of calculation of plasticity index of two-process surfaces was proposed [16]. However, this work was restricted only to isotropic textures.

The main aim of this work is to develop method of plasticity index calculation for honed and particularly plateau honed cylinder surfaces. This would help to better predict the ability of cylinder to plastic deformation and hence wear. The other aim is to find if the calculation of the plasticity index by taking the cylinder axis and circumference as the principal axes leads to tolerable errors.

IOP Conf. Series: Journal of Physics: Conf. Series **1183** (2019) 012003 doi:10.1088/1742-6596/1183/1/012003

2. Materials and methods

Twenty five honed (one-process) and plateau honed (two-process) cylinder liner surface topographies were measured by white light interferometer Talysurf CCI Lite (1024 x 1024 points, sampling intervals 3.3 μm in perpendicular directions, measurement area 3.28 x 3.28 mm). Honing angle α (see figure 1) of one process surfaces was in the range: 15-70^0, but of two-process textures 50-60^0. The height of one-process textures, determined by the Sq parameter (the root-mean-square value of the surface deviations) was between 0.5 and 2.8 μm. The values of the Sq parameter of two-process textures were smaller, up to 1 μm. These surfaces are better characterized by the root-mean-square value of the surface deviations Spq [17]. This parameter was in the following range: 0.1 – 0.4 μm. The form was removed by a polynomial of the 2nd degree. Digital filtration was not used.

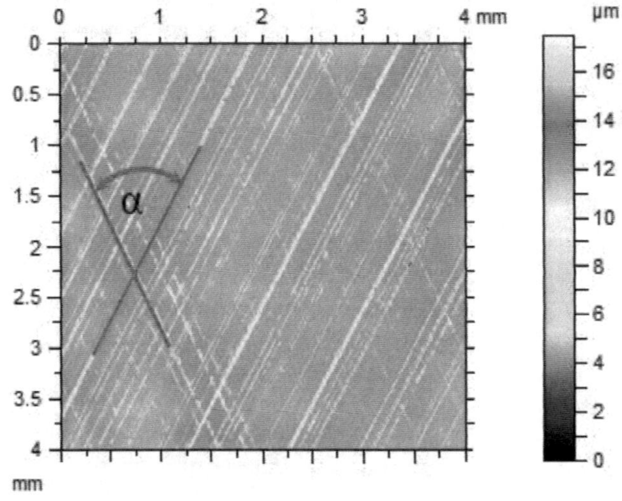

Figure 1. Honing angle α

In both analyzed methods the plasticity index was calculated using equation (2). The equivalent radius of curvature in the traditional method was calculated in the axial and circumferential directions, while in the proposed method for principal curvature radii of individual asperities.

For measured surfaces, summits were identified on the basis of their eight neighbors [5, 18], which means that a surface point was classified as a summit when its ordinate was higher than ordinates of eights neighboring points. For each summit, the ratio of the smaller and the larger radii of curvature in the axial and circumferential directions γ_1 was computed using a three-point Lagrangian formula [19]. Then the ratios of the radii in orthogonal directions for different orientations (9^0 resolution) were computed and the smallest value, γ_2 was recorded. Average values of these ratios were stored.

The error in the plasticity index estimation using γ_1 (traditional method) rather than the true value γ_2 (proposed method) was computed. Taking the sum of any two orthogonal curvatures as equal to the sum of the principal curvatures [4] leads to an error estimate of:

$$\Delta = \left(\frac{1+\gamma_2}{1+\gamma_1}\right)^{0.5} \left(\frac{\gamma_1}{\gamma_2}\right)^{0.25} -1 \qquad (3)$$

The method of calculation of plasticity index of two-process surfaces is proposed in Reference [16]. In this approach, only the summits which taking part in contact were taken into consideration.

3. Results and discussions

One-process surfaces

Hybrid and spatial properties of surfaces and hence the values of plasticity index depend on the sampling interval [5, 20, 21, 22]. The smallest sampling interval was 3.3 μm, while the sampling intervals used in calculating various radii varied with angular orientation, but were always the same in the orthogonal directions. The γ_1 values were larger than 0.3; in most cases the γ_2 values were similar or greater than 0.2, so the correction factor [8] is of little significance. Similar results were obtained previously [23]. Figure 1 presents an example of the analysed cylinders with values of the degrees of anisotropy γ_1 and γ_2 as well as of errors of the plasticity index calculations Δ.

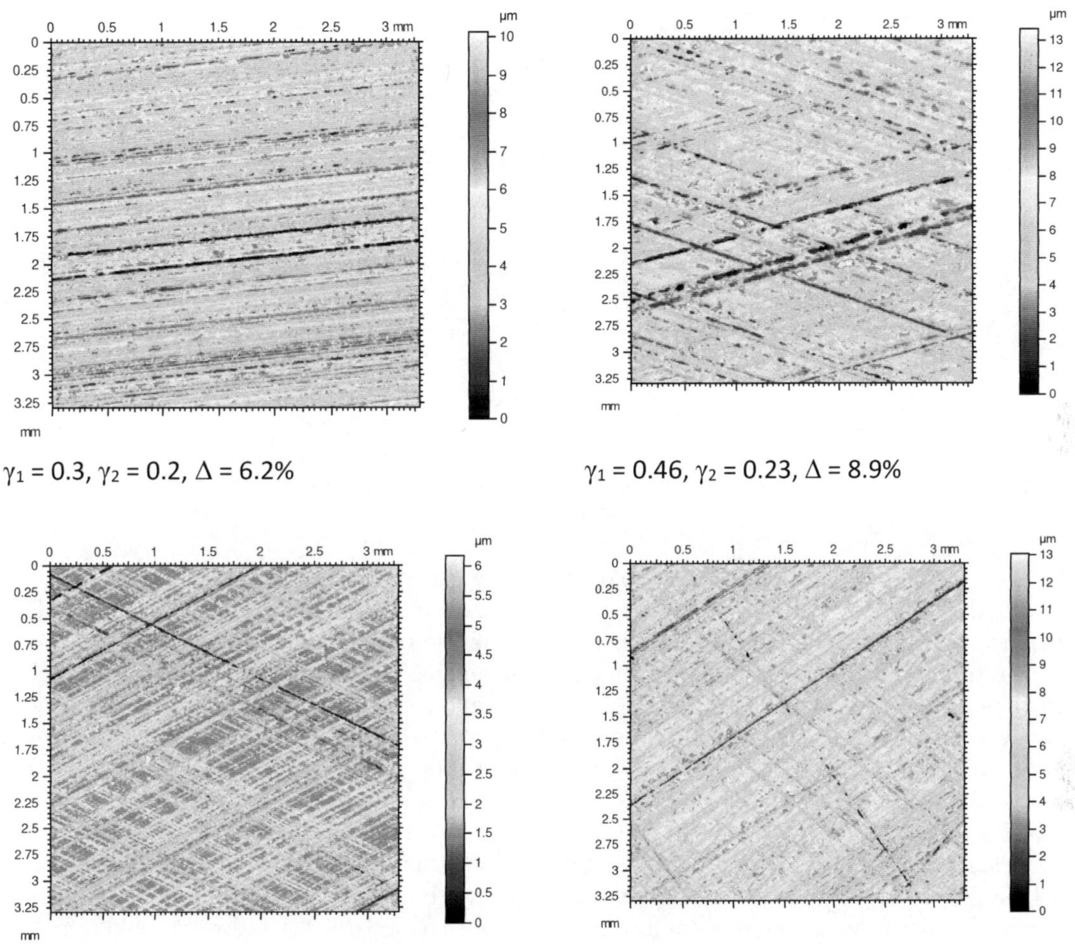

$\gamma_1 = 0.3$, $\gamma_2 = 0.2$, $\Delta = 6.2\%$ $\gamma_1 = 0.46$, $\gamma_2 = 0.23$, $\Delta = 8.9\%$

$\gamma_1 = 0.48$, $\gamma_2 = 0.2$, $\Delta = 12.2\%$ $\gamma_1 = 0.57$, $\gamma_2 = 0.2$, $\Delta = 13.5\%$

Figure 2. Contour plots of one-process measured cylinder liners with values of γ_1, γ_2 and Δ.

The errors of the plasticity index estimation using γ_1 than γ_2 led to an overestimation of the plasticity index. These errors (equation 2) were between 6 and 15.5%. Such errors are reasonably acceptable bearing into mind the large uncertainties in the measurement of some surface topography parameters and the assumption that circumference and axis define the major axes suffices. However, it is not difficult to compute more correct γ_2 instead of γ_1. From the analysis of cylinder liners contour plots, one can assume that the errors should be larger for larger honing angles. These assumptions were confirmed (see figure 3).

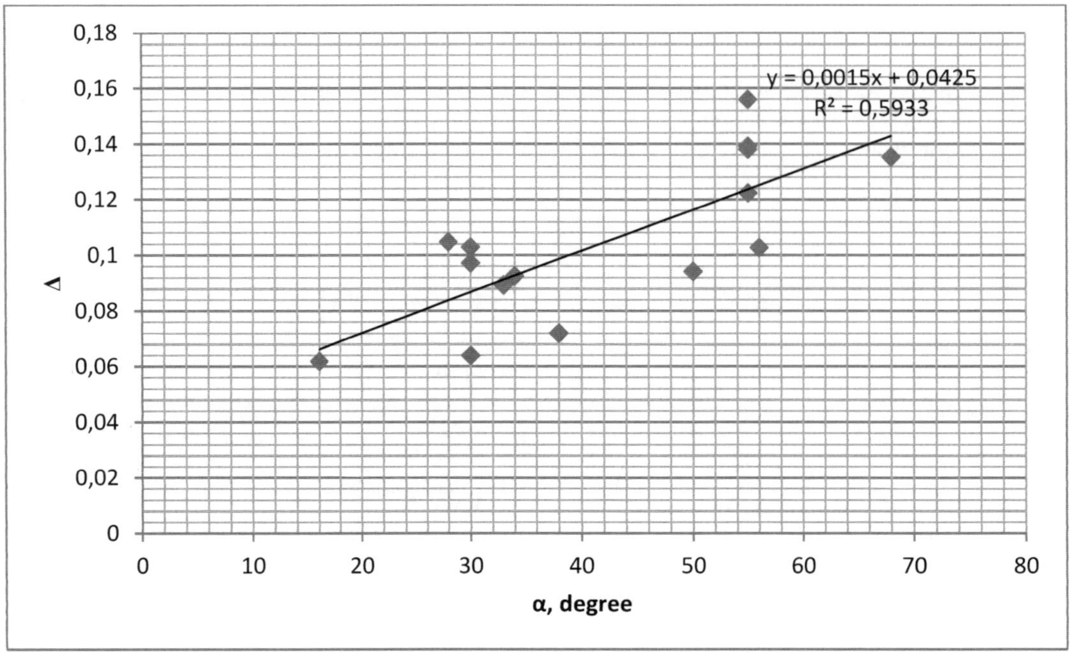

Figure 3. Dependence between honing angle and error of plasticity index calculation

Two-process surfaces

Calculation of the plasticity index of the whole surface by can seriously overestimate its value. In order to compute the plasticity index of the two-process surface in a correct way, the special procedure was developed. All summits should be recognized. The heights of all summits should be stored. Then the special profile containing heights of asperities was obtained. This profile was analysed using probability approach [9, 10, 11, 12, 13], in which material ratio is expressed as Gaussian probability in standard deviation (s) values, plotted linearly on the horizontal axis (-3s = 0.13%, -2s = 2.28%, -s = 15.8%, 0 = 50%, s = 84.13%, 2s =97.72%, 3s = 99.87%). For profiles composed of two Gaussian distributions, the material probability curve exhibits two linear regions. There are two parts of the profile obtained from heights of summits – plateau and valley. There are 3 kinds of summits on the two-process surface (see figure 4). For summits of type 1, both the summit and its neighbouring points belong to the plateau part PP. A summit of type 2 belongs to the plateau part, while its neighbouring points to valley part VP. However, a summit of type 3 belongs entirely to the valley part. To calculate the standard deviation of asperity height and mean radius of summits curvature only summits of types 1 and 2 should be analysed. Therefore only summits positioned vertically between the highest summit and the horizontal line visible in figure 4 were taken into consideration. The procedure of the plasticity index calculation for two-process surfaces is detailed described in Reference [16].

IOP Conf. Series: Journal of Physics: Conf. Series **1183** (2019) 012003 doi:10.1088/1742-6596/1183/1/012003

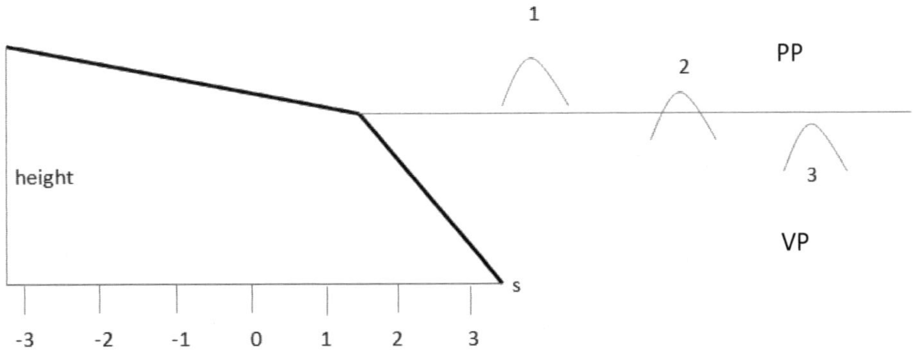

Figure 4. Three kinds of summits

For identified summits, degrees of anisotropy γ_1 and γ_2 were computed.The anisotropy ratios were calculated for all summits and only for summits belonging to plateau parts. The analysis of only summits of types 1 and 2 led to an increase in the degree of anisotropy γ_2 calculated for all summits. Smaller but similar changes in calculating the γ_1 ratio occurred. In all analysed cases the γ_2 values were greater than 0.2, so it is not necessary to use the correction factor $\varphi(\gamma)$. However, after study of all summits, this correction would be needed. The errors of the plasticity index calculation were between 10.5 and 11.9%. Similar values of the honing angle may be the most probable reason of stable errors. Figure 5 shows an example of the analysed cylinders with values of the degrees of anisotropy γ_1 and γ_2 as well as of errors of the plasticity index calculations Δ.

However, the most important difference in plasticity index calculation for all summits and summits belonging to the plateau part is the standard deviation of asperity height σ_s. This parameter is much higher after considering all summits existed on the surface, leading to overestimation of the plasticity index.

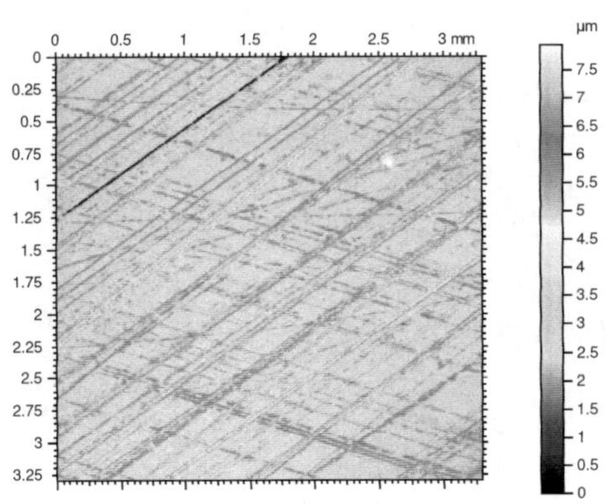

$\gamma_1 = 0.53$, $\gamma_2 = 0.21$, $\Delta = 11.8\%$

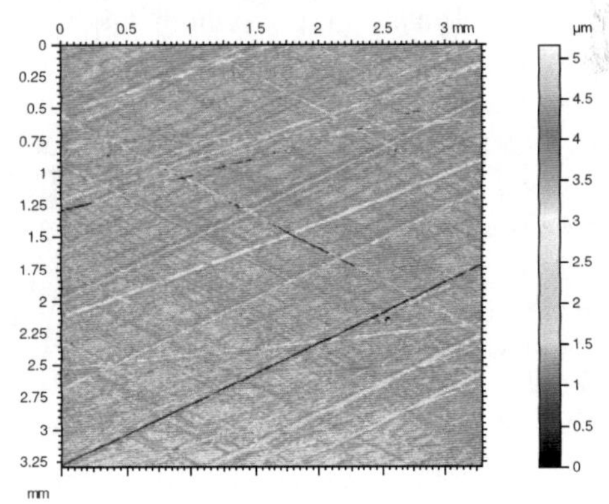

$\gamma_1 = 0.51$, $\gamma_2 = 0.21$, $\Delta = 11.4\%$

IOP Conf. Series: Journal of Physics: Conf. Series **1183** (2019) 012003 doi:10.1088/1742-6596/1183/1/012003

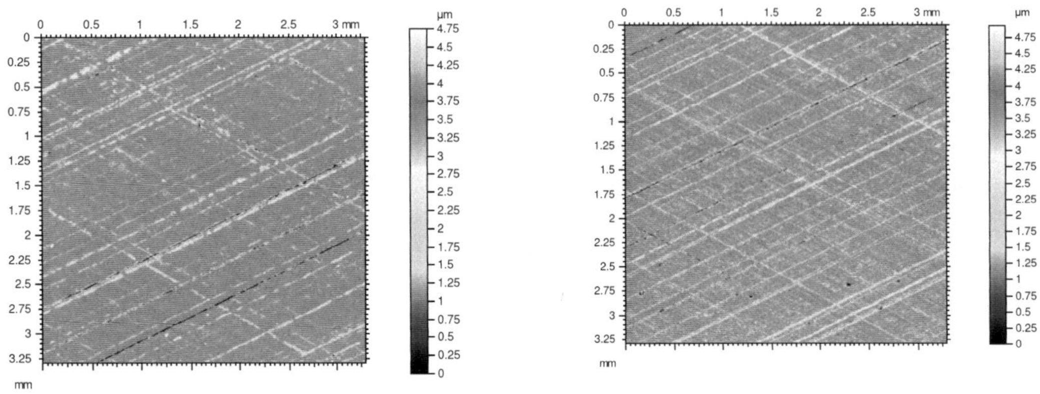

$\gamma_1 = 0.55$, $\gamma_2 = 0.22$, $\Delta = 11.2\%$ $\gamma_1 = 0.22$, $\gamma_2 = 0.5$, $\Delta = 10.9\%$

Figure 5. Contour plots of two-process measured cylinder liners with values of γ_1, γ_2 and Δ

Greenwood and Wiliamson [1] in 1966 developed a basic contact model of isotropic surface, in which all asperity summits have the same radius but their height varied randomly. The interaction of asperities was not taken into consideration. From this time, new better models were developed. For example, the Boundary element method-BEM in a half space is well established and used method for contact simulations. Also BEM does not have problems with the definition of a summit. However, no alternatives to plasticity index have not been developed yet.

4. Conclusions

Only a tolerable level of error is introduced in plasticity index calculations by taking the cylinder axis and circumference as the principal axes when the honing angle of one-process textures is between 15 and 70^0. The errors are larger when the honing angles are larger. However, a procedure for calculating the anisotropy degree in better way can be easily developed. Similar and even smaller errors were obtained when two-process cylinder liners are analysed for the honing angles between 50 and 60^0. The analysis of only summits which might take parting contact led to an increase in the degree of anisotropy γ_2 calculated for all surface summits.

It is not necessary to use the correction factor for plasticity index calculation of honed and plateau-honed cylinder surfaces because the correction factor γ_2 was typically higher than 0.2.

The errors of the plasticity index estimation calculating the equivalent radius of curvature in the axial and circumferential directions led to an overestimation of the plasticity index.

References

[1] Greenwood JA and Williamson JBP 1966 Contact of nominally flat surfaces. *Proc. Roy. Soc. (London)* **A295** 300-319

[2] Hirst W and Hollander AE 1974 Surface finish and damage in sliding. *Proc. Roy. Soc., London* **337** 379-394

[3] Poon CY and Sayles RS 1992 The classification of rough surface contacts in relation to tribology. *J. Phys. D: Appl. Phys.* **25** 249-256

[4] Nayak PR 1973 Some aspects of surface roughness measurement *Wear* **26** 165-174

[5] Nayak P R 1971 Random process model of rough surfaces *ASME Journal of Lubrication Technology* **93** 398-407

[6] Pawlus P and Zelasko W 2012 The importance of sampling intervals in contact mechanics. *Wear* **276** 121-129

[7] Bush A W, Gibson R D and Keogh G P 1979 Strongly anisotropic rough *surface ASME Journal of Tribology* **1010** 15-20

IOP Conf. Series: Journal of Physics: Conf. Series **1183** (2019) 012003 doi:10.1088/1742-6596/1183/1/012003

[8] Wu C, Zheng L 1988 A general expression for plasticity index *Wear* **121** 161-172
[9] Malburg MC, Raja J. 1993 Characterization of surface texture generated by plateau-honing process. *CIRP Annals* **42** 637-640
[10] Zipin R B 1983 The analysis of profile strata for surface texture specification *Appl. Surface Science* **15** 334-360
[11] Whitehouse DJ 1985 Assessment of surface finish profiles produced by multi-process manufacture. *Proceeding of the Inst. Mech. Engrs*. **199** 263-270
[12] Sanna Reddy H, Raja J and Chen K 1997 Characterization of surface texture generated by multi-process manufacture. *Transactions of the 7th International Conference on Metrology and Properties of Engineering Surfaces, Gothenburg* 111-118
[13] Pawlus P and Grabon W 2008 The method of truncation parameters measurement from material ratio curve. *Precision Engineering* **32** 342-347
[14] Codgell JD 2008 A convolved multi-Gaussian probability distribution for surface topography application. *Precision Engineering* **32** 34-46
[15] Leefe SE 1998 "Bi-Gaussian" representation of worn surface topography in elastic contact problems *Tribology for Energy Conservation* 281-290
[16] Pawlus P, Zelasko W, Reizer R, Wieczorowski M 2017 Calculation of plasticity index of two-process surfaces *Proceedings of the Institution of Mechanical Engineers, Part J: Journal of Engineering Tribology* **231** 572-582
[17] Leach R 2013 Characterisation of Areal Surface Texture *Springer-Verlag, Berlin, Heidelberg*
[18] Sayles R Thomas T. 1979 Measurements of the statistical properties of engineered surfaces. Transaction of ASME, Journal f Lubrication Technology **101** 409-417
[19] Whitehouse D.J. 1978 The digital measurement of peak parameters on surface profiles. *J. Mech. Eng. Sci.* **20** 221-227
[20] Thomas TR and Rosen BG 2000 Determination of the optimum sampling interval for rough contact mechanics. *Tribology International* **33** 601-610
[21] Pawar G, Pawlus P, Etsion I et al. 2013 The effect of determining topography parameters on analyzing elastic contact between isotropic rough surfaces *ASME Journal of Tribology* **135/1** No. 011401
[22] Pawlus P, Chetwynd DG 1996 Efficient characterization of surface topography in cylinder bores *Precision engineering* **19 (2-3)** 164-174
[23] Pawlus P, Chetwynd DG 1996 3-D surface texture characterisation of cylinder bores *IX International Colloquium on Surfaces* 473-484

IOP Conf. Series: Journal of Physics: Conf. Series **1183** (2019) 012004 doi:10.1088/1742-6596/1183/1/012004

Focus variation measurement and advanced analysis of volumetric loss at the femoral head taper interface of retrieved modular replacement hips in replica

Karl Walton[1], Maila Petrucci[2], Radu Racasan[1], Liam Blunt[1], Alister Hart[3], Paul Bills[1]

[1]EPSRC Centre for Innovative Manufacturing in Advanced Metrology, University of Huddersfield, UK

[2]Università di Bologna - Via Zamboni, 33 - 40126 Bologna

[3]London Implant Retrieval Centre, University College London, UK

E-mail: k.walton@hud.ac.uk

Keywords: Replication, hip, retrieved, wear, corrosion, focus variation, areal, volume loss

Abstract. This paper offers a technique for non-contact assessment of material volume loss at the femoral head taper interface of modular replacement hips with novel use of the focus variation principle. A novel 3D to areal data conversion technique also allows powerful areal data analysis techniques to be applied to point cloud (3D) measurements for volume loss optimisation. Accurate characterisation of retrieved femoral head tapers is important to assess their in vivo performance particularly with respect to the bio toxicity of wear and corrosion products.

A cohort of 8 retrieved femoral heads showing a wide range of degradation ($0.5mm^3$ < volume loss < $12mm^3$) [1] was selected for the development of this technique. Using this common cohort of retrieved hips, the current technique was benchmarked against the well proven roundness measurement machine (RMM) method. This existing technique generates areal (2.5D) data and exploits a range of existing areal analysis techniques to optimise volume loss assessment. For benchmarking continuity and to exploit the same areal techniques, volume loss analysis for the current technique was carried out using the software written for the existing RMM method.

The focus variation instrument's integrated language was used to write script to convert (un-wrap) the taper surface 3D data into areal format.

The current method shows a mean absolute difference in volume loss of 14% (-12% signed) from that of the benchmark with a range of 1% to 27%. The spread of measured values is significantly higher for the current method than for the benchmark. However, it is noted that replication can offer the advantage of capturing the whole taper surface on some taper types where physical access is limited for a stylus based roundness method.

The current technique is also compared to the existing Redlux™ technique in which replicated female tapers are measured using a confocal instrument. The current technique is shown to have comparable performance to the Redlux™ technique but offers a more sophisticated methodology for volume loss analysis. In addition the current technique offers new instrumentation and analysis tools to the field.

IOP Conf. Series: Journal of Physics: Conf. Series **1183** (2019) 012004 doi:10.1088/1742-6596/1183/1/012004

Small uncontrolled casting variations are noted in the current technique, resulting in poorer performance with small volume loss samples where the influence of this effect is most pronounced. However, given the simplified assumptions of the volume loss calculation where results may be skewed by deposits, some uncertainty will be evident with any approach.

1. Introduction

Learmonth, Young and Rorabeck [1] refer to the total hip replacement procedure as the operation of the century. Within England and Wales more than 700,000 such primary procedures were carried out between April 2003 and December 2014 with more than 71,000 in 2012, with osteoarthritis being reported as the primary diagnosis in over 90% of cases [2].

For hip resurfacing or total hip replacement designs several combinations or bearing surface material couples are used; metal or ceramic head on ceramic, metal, ceramic or polyethylene acetabular cup. Increasingly popular is the modular system of total hip replacement, Figure 1 shows the components and taper contact surfaces of a typical modular hip replacement. The modular taper interface design allows individual components to be selected to allow joint geometry to be tailored to individual recipients during surgery. However, these taper interface surfaces lack fully developed standardisation and tolerancing for; length, diameter at depth, cone angle, roundness, straightness and surface finish. Several studies have reported significant wear and the formation of corrosion products at these taper interfaces [3-5] see Figure 2. Pseudotumours caused by wear and corrosion products are commonly seen in joint capsules during revision surgery [4, 6]. The current work focuses on the "large head metal on metal" (LHMoM) design which are more prone to taper corrosion and wear due to increased bending moments at the stem/head taper interface. Thought the taper interfaces of all modular systems are susceptible and the techniques and analysis detailed in the current work are applicable. A number of mechanisms have been suggested to explain the ionic contamination and metallic debris. The use of insufficient force or simple misalignment during joint assembly can result in subsequent wear from micro-motion. Femoral head and stem materials are typically dissimilar and this can result in electrochemical corrosion due to the resulting galvanic cell [7]. Goldberg et al [8] detail mechanically-assisted crevice corrosion, and the similar mechanism of fretting corrosion is offered in [9].

Quantifying taper interface material loss in recovered implants is central to the analysis of hip failure [10]. Though it should be noted there is no uniform correlation between material loss and any adverse response in the recipient as sensitivity to metal ions varies widely between individuals. A number of methods have been used to measure this taper interface volume loss;

IOP Conf. Series: Journal of Physics: Conf. Series **1183** (2019) 012004 doi:10.1088/1742-6596/1183/1/012004

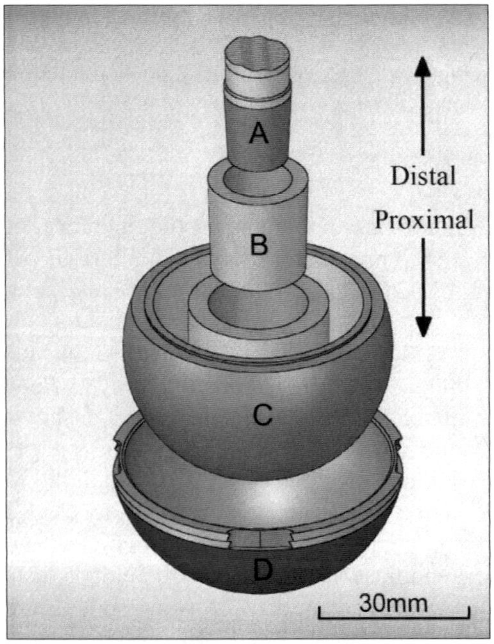

Figure 1. Metal-on-metal hip replacement components and taper interfaces, showing; A) Femoral stem trunnion (sectioned), B) Modular taper adapter sleeve, C) Modular head, D) Acetabular cup.

Figure 2. Retrieved metal-on-metal femoral head showing corrosion and wear on internal taper surface (un-sleeved type).

coordinate measuring machines (CMM) are capable as detailed in [11], though they have limited resolution in the plane perpendicular to the sample surface (vertical resolution) when compared to other techniques detailed in the current work. Roundness Measurement Machines (RMM) such as the Talyrond 365TM (Taylor Hobson, Leicester UK) [12, 13] offer the vertical resolution (30nm) and stylus tip radius (5μm) of surface profilometers and thus better accuracy than a CMM for this application. Redlux™ (Redlux Ltd, Southampton UK) have previously reported a method [14, 15] similar to that detailed in the current work where an optical instrument is used to measure taper interfaces in replica. Most recently Redlux™ have developed a non-contact optical instrument (Taperlux™) for the direct measurement of internal tapers, thought the authors are unaware of any peer reviewed publications or data using this instrument.

All existing methods for calculating volume loss at taper interfaces are subject to two inherent limitations;

No prior information about the unworn surface of the taper is available and thus nominally "unworn" region of the retrieved taper must be relied upon to establish a datum from which to calculate volume loss. Typically there are two such unworn taper regions, one distal and proximal, though the accuracy of a datum based on these can be compromised by surface roughness or even by one region being absent.

A simplified approach is taken to quantify volume loss as both material deposition and removal can occur at the interface. Thus volume loss is considered to be equivalent to the material void between the estimated position of the unworn (original) taper surface and the measured surface of the retrieved taper. Material deposited on a worn region will thus be considered as taper material and not be reported as lost material.

The method detailed in the current work involves casting elastomer replicas of femoral head tapers and using a focus variation (FV) instrument (Alicona IFM G4) to characterise the surface volume loss. The nominal characteristics of the focus variation (FV) instrument group are detailed in ISO

IOP Conf. Series: Journal of Physics: Conf. Series **1183** (2019) 012004 doi:10.1088/1742-6596/1183/1/012004

25178-606 [16] and its operating principles and application are detailed in [17-19] **Figure *4*** shows the FV instrument and rotary stage arrangement used to measure the taper replicas. The FV instrument operates by scanning through the (z) direction envelope of surface heights. The sharpness of the imaged surface texture at the sensor array for each (x,y) location in the field of view is a measure of focus precision and thus surface point heights are assigned. This approach is typically used to capture point height data for 'areal' surface topography characterisation (2.5 dimensional data). In addition, 3D data can also be captured, for example from an axisymmetric component with suitable mounting in a forth axis rotary stage.

Acceptance and performance re-verification standards for coordinate measurement systems employing optical distance sensors and rotary axes are detailed in ISO 10360-8 / 3 [20, 21] respectively. However, with the increased use of optical instrument in coordinate measurement type roles it is clear that there are significant gaps in CMM standards pertaining to "optical distance sensors" and particularly with respect to FV instruments. Unlike the Redlux™ confocal technique, FV is not explicitly included in the scope of [20] nor does the FV instrument used in the current work have the working volume requirements to employ the test artefacts from that standard. In addition little consideration has been given to the influence of image field stitching processes on coordinate measurement. Sun and Claverley [22] reported standard format probing errors for their own data reduction (sampling) methods for the Alicona FV instrument. Importantly they note that to ensure consistency in their results outliers were removed from their surface data by filtering even though user defined filtering of data is explicitly excluded from testing in [20]. An investigation into the simultaneous 4 axis performance of the Alicona FV instrument was carried out by Moroni, Syam, and Petro [23]. This study reports positional errors as large as 500µm when measuring test sphere positions on a heavy artefact across the instruments whole working volume, involving the simultaneous use of all 4 instrument axes. Given that only 3 axes are used independently over a small measurement range with a light sample in the current application, errors of this reported scale will not be seen. The aim of the current work is to develop a technique for the assessment of taper material loss in replica using a focus variation instrument and apply existing advanced areal volume loss analysis software. Then benchmark the technique against the existing well proven RMM technique in [12] by applying both techniques to the same group of retrieved head tapers. To apply the volume loss analysis software from [12] to the current results, script was written to process the current data and change its format. The current technique is then compared to the similar existing replication technique due to Redlux™ [14, 15], though a different groups of tapers is used.

2. Methods and materials

To replicate the tapers Microset™ fluid grade 101FF grey was used with working time of 4 minutes and curing time of 30 minutes and with quoted resolution and shrinkage values of 0.1µm and less than 0.1% respectively. The taper replication was carried out using the alignment and replication fixture shown in Figure 3 this fixture (D) ensured that the taper replica mounting tube (A) and the head (C) were coaxial. The head was aligned using the square end of the mounting tube and replication compound was injected down the mounting tube to fill the; taper, extension sleeve (B) and lower mounting tube end. The compound was then allowed to cure and extracted from the head by way of the mounting tube. All of the recovered heads in the current cohort were of the design seen in Figure 1 with 12/14mm taper trunnion (A) fitting into a 12/14mm //18/20mm taper adapter sleeve (B) which remained in situ (after retrieval) in a 18/20mm head taper. Thus a gap remains between the proximal (bottom) end of the sleeve and the taper socket in the head, such a re-entrant void complicates the extraction of the replica. Consequently the voids were filled with a suitable non-curing elastomer that could later be recovered. In addition the taper casting void was extended at both ends, proximally into the non-curing elastomer and distally by way of the extension sleeve (Figure 3 (B)) this is discussed further in section 3. Figure 4 shows a replica taper and mounting tube in the FV instrument rotary stage collet (4th axis) and evidence of the casting extensions can be seen.

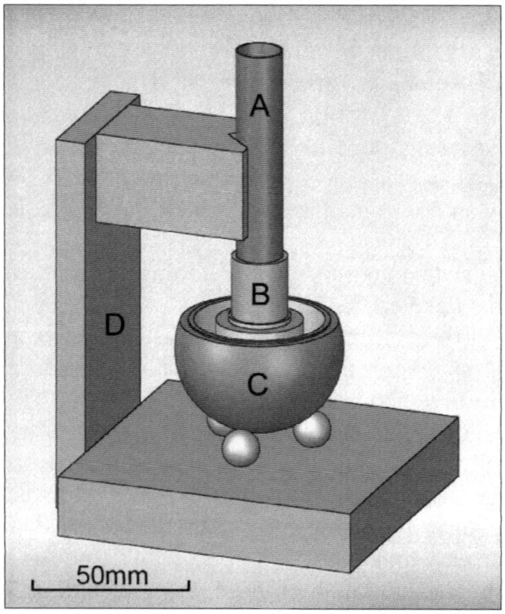

Figure 3. Showing alignment and replication fixture; A) Mounting tube, B) Replication extension sleeve, C) Large MoM femoral head, D) Three sphere alignment and replication fixture.

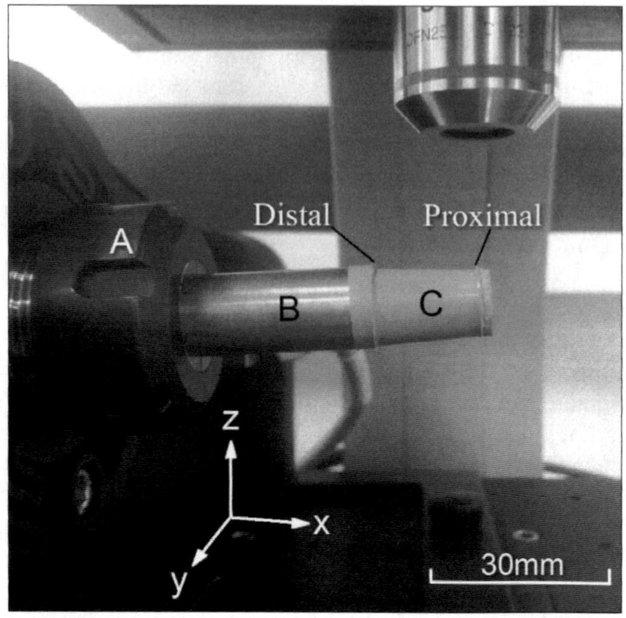

Figure 4. Showing a replica taper mounted in the instrument rotary stage collet in the measurement position. A) Instrument rotary stage, B) Replica taper mounting tube, C) Replica taper (MicrosetTM).

The instrument scans through a (z) direction envelope to capture individual fields the height of this envelope is largely determined by the curvature of the taper within the field. A series of fields are then captured along the axial length of the component, these are stitched together with 10% overlap to form a single longitudinal field. The rotary stage then increments about its precision calibrated axis to capture the next longitudinal field, the incremented angle is determined from the overlap of the longitudinal fields specified by the user (default 30%). Stitching alignment is calculated based on the; overlapping region surface point heights, point true colour information and relationship to the stage axis. Thus measurement precision and extent of overlap dictate stitching accuracy.

To minimise the (z) envelope height and thus the measurement duration the rotary stage axis was adjusted to a suitable angle (approximately 2.5°) such that the measured surface of the taper replica was perpendicular to the (z) direction. The rotational axis of the stage is only referenced for field stitching and for no other purpose. Thus any misalignment of the taper replica axis and its mounting tube only serve to increase the measurement duration and have no influence on the accuracy of the subsequent volume analysis.

Measurement duration, resolution and file size considerations dictated the use of the instruments x5 objective lens with a vertical resolution of 0.5µm. This vertical resolution value was verified by its optimisation of measured data quality (repeatability) [19] and the lack of any clear evidence of improved volume loss resolution over the effective vertical resolution range of the 5x lens (0.1 to 2µm). Due to final file size limitations points are spaced at 14µm in both the x and y directions thought this is down sampled from a native optical resolution for this configuration of 2.2 µm. For each taper measurement 8 axial measurement fields were stitched to form each of 26 longitudinal measurement regions each having approximately 2 million points. During stitching to form the taper point cloud the overall number of points was reduced by a factor of 10 to give a practical maximum for the system of approximately 4.5 million. Measurement duration varied between 1.5 and 3 hours (in temperature controlled conditions) depending on the extent of the axial runout resulting from replication fixturing misalignment.

Volume loss analysis was carried out using the Matlab™ (The Mathworks Inc., Natick, USA) software developed in [12]. In all other techniques volume loss determination is carried out in a 3 dimensional format, with the unworn taper surface being considered and fitted as a simple conic primitive. However the current software uses the more sophisticated approach of fitting the unworn taper surface as a plane to the measured taper data in areal (planar) format. This approach has the benefit of allowing the removal of potential manufacturing form errors from the retrieved taper by polynomial fitting in both the longitudinal ("hogging" or "sagging") and radial directions. These options are not available with the data in 3D format and thus these errors would influence the volume loss calculation. The datum plane of the unworn taper surface is then fitted to manually selected unworn taper regions and a data thresholding approach based on the Abbot-Firestone curve [24, 25] (see [12] further details) is used in place of a simple volume subtraction. However this approach requires that the measured taper point cloud (3D) data be unwrapped and transformed to point height (areal) format.

Script was written using Alicona's integrated language to achieve this unwrapping and format transformation. The script processing involved; axial alignment of the point clouds by fitting a conic primitive to unworn regions of the data set, perpendicular trimming of the point cloud ends to match the extent of each sample's taper surface measured by the RMM technique for, "unwrapping" of the point cloud to a point height format for final volume loss analysis. The "unwrapping" process was achieved by extracting longitudinal taper surface profiles at intersections of the taper point cloud (at 1° rotational increments) and a plane passing through the tapers rotational axis. The resulting profiles were resampled to achieve the same uniform '14µm' sample spacing as the original measurements, giving approximately 0.5 million points in each point height file. Volume loss analysis was carried out on two sets of the resulting point height data for the samples; one set was filtered using a spline S filter with a nesting index [26] of 50µm and the other had no filter applied. Unworn regions of the taper surface map (see **Figure** *8* (1)(u)) were then selected in the Matlab™ software [12], automated polynomial form fitting to these regions then allowed removal of manufacturing form errors in the taper. The taper surface map was then levelled with respect to a plane fitted (linear least square) to the unworn regions (fitting can also be polynomial). The taper surface map was then thresholded (**Figure** *8* (2)) to the "zero" height level of the cumulative surface height plot (Abbot-Firestone curve **Figure** *8* (3)) the zero height having been defined by the unworn region selection in the previous step. This thresholding delineates the surface debris above and defines the volume loss below the surface of the nominally unworn original taper.

In conventional use the; selection of unworn regions, thresholding process and determination of polynomial form all require experience and judgement (but show good interoperability [12]). Therefore, in the current study the specified default selections are made in all cases to limit bias. Thus it is possible that better agreement with the benchmark results in [12] could have been achieved if the current processing had been carried out by an experienced operator.

3. Results and discussion

3.1. Benchmarking against the existing RMM method

Table 1 shows the volume loss data for the retrieved femoral head tapers acquired by both the RMM [12] and the FV method detailed in the current work. Replication and measurement of samples 3 and 6 was repeated four times and the mean values of volume loss, standard deviations and coefficients of variation are shown and included in the main volume difference analysis. The FV volume loss data is expressed with respect to the RMM data in both mm^3 and percentage terms and 'absolute' mean values are quoted. For comparison volume difference values are given for the FV in both filtered and unfiltered form. Repeatability data is also given for the RMM technique, thought this data is from a sample not measured in the current study and has a lower volume loss than the samples reported for the current study.

IOP Conf. Series: Journal of Physics: Conf. Series **1183** (2019) 012004 doi:10.1088/1742-6596/1183/1/012004

Table 1. Volume loss data for the FV and RMM technique including repeatability data with mean, standard deviation (σ) and coefficient of variation (Cv) values. Cell colouring indicates where mean repeatability values are used.

			FV					
Sample	**RMM**		No Filter			**50µm Filter**		
	Vol		Vol	Diff	Diff	**Vol**	**Diff**	**Diff**
	mm³		**mm³**	**mm³**	%	**mm³**	**mm³**	%
1	**0.23**		0.22	0.01	-5.15	0.17	0.05	-23.70
2	**0.78**		0.54	0.24	-30.92	0.63	0.15	-19.39
3	**1.93**		1.91	0.01	-0.77	**1.91**	0.02	-0.93
4	**2.39**		1.79	0.60	-25.21	1.83	0.56	-23.49
5	**2.89**		2.29	0.60	-20.79	2.27	0.62	-21.47
6	**4.99**		5.22	-0.23	4.69	**5.17**	-0.19	3.75
7	**7.41**		5.84	1.57	-21.18	6.03	1.39	-18.72
8	**11.67**		12.38	-0.71	6.10	12.21	-0.54	4.62
			mean	0.50	14.16	**mean**	**0.44**	**14.28**

Repeatability						
Sample	**Volume loss (mm³)**					*σ/mean*
	range			**mean**	**σ**	**Cv**
	No Filter					
3	1.76	to	2.06	1.91	0.126	0.066
6	5.00	to	5.36	5.22	0.154	0.030
	50µm Filter					
3	**1.76**	**to**	**2.07**	**1.91**	**0.131**	**0.069**
6	**4.85**	**to**	**5.35**	**5.17**	**0.221**	**0.043**
RMM	**0.60**	**to**	**0.69**	**0.64**	**0.023**	**0.036**

The RMM technique [12] captured 180 axial taper profiles with 3500 points in each (total 630,000 points). This same study, and that due to Underwood et al [13] also show the insensitivity of the volumetric analysis to the number of longitudinal profiles captured. As a consequence of the acquisition technique in [12] the taper surface data is in the form of a point height map (.sur) instead of the more conventional CMM point cloud format.

It is noted that surface point height distribution of the FV technique is subject to a greater level of noise including outliers than is the RMM technique. This observation is in line with that expected for the comparison of surface topography data acquired by contact stylus and focus variation instruments. This raises the question of applying a noise (S) filter [26] to the FV data to offset the influence of the increased magnitude of this high frequency band in the FV data. For optical CMM little guidance is available in this respect and [20] prohibited user specified filter settings for re-verification purposes, but in practice filtering "noise " from the FV data is a logical step and Table 1 shows the comparison of the filtered and unfiltered data. On average a 50µm noise filter produced only a limited effect on the volume loss difference values. However, the effect on samples 1 and 2, those with smallest volume loss is significant and it is clear that this noise frequency band can play an important role. This strongly indicates that the use of such a filter is appropriate. Thus the volume loss difference values calculated from the filtered data will be reported in the current work. An absolute mean difference in volume loss from the RMM values of approximately 0.4mm^3 or 14% (-12% signed) is seen for the whole sample group with ranges of 0.02 to 1.39mm^3 and 0.93 to 23.7% respectively. Coefficients of variation (standard deviation/mean) values show the two techniques to have similar variability. However, it is considered likely that the RMM technique would maintain similar absolute variation values over larger volume loss samples and thus show better repeatability in general than the current technique.

Figure 5 shows the difference in volume loss from the RMM technique for the FV technique in % terms, and the samples have been numbered in order of increasing volume loss. The sample results are in two distinct groups, those with minimal difference in volume loss and those with a greater difference. Importantly, where the difference is appreciable it is negative and the replica technique is reporting lower volume loss than the RMM technique by approximately 20%. In general it is noted that the replication in this group of samples is subject to the type of form defect discussed and illustrated with respect to Figure 9. In addition it could be argued that with the exception of samples 3 and 7 that a trend is apparent where low volume loss in the sample is associated with higher measured difference in volume loss. This result would suggest that the resolution of the current technique is such that its accuracy decreases over the range of volume loss magnitudes seen in the sample.

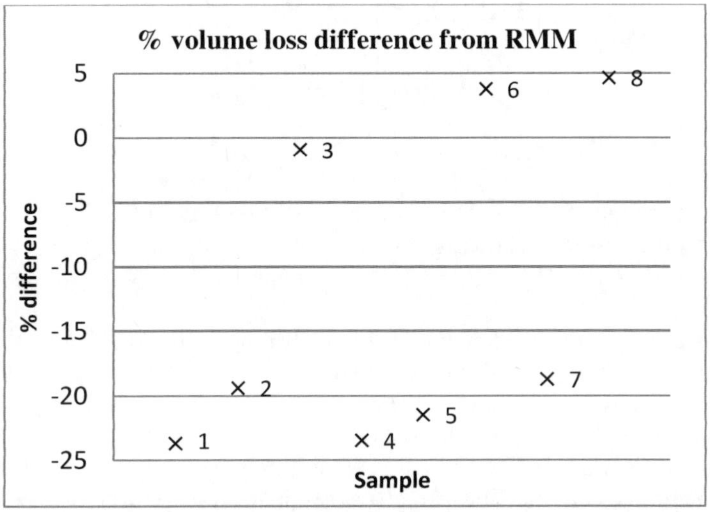

Figure 5. Plot of sample number with % volume loss difference from RMM

Figure 6 shows the deviation map representation of the point cloud FV measurement from sample 5. The false colour scale indicates the deviation from a conic primitive fitted to the unworn regions of the taper as seen in **Figure 8** (u). **Figure 6** shows a number of features of interest A) to E); (A) shows the fine pitch tooling marks of the original (unworn) femoral taper sleeve surface (**Figure 1** (B)). (B) Shows a degree of corrosion on the exposed distal surface of the taper where it has been in contact with bodily fluids adjacent to the taper contact region. (C) Shows the approximate distal and proximal extent of the contact region of the femoral stem trunnion and the sleeve taper surface (**Figure 1** (A) and (B) respectively). (E) shows the somewhat more course pitch of the tooling marks of the trunnion surface having been transferred from that surface during service. The current technique resolves theses characteristic features of retrieved taper surfaces that are also noted in studies using other measurement techniques both directly and in replica [13, 14].

IOP Conf. Series: Journal of Physics: Conf. Series **1183** (2019) 012004 doi:10.1088/1742-6596/1183/1/012004

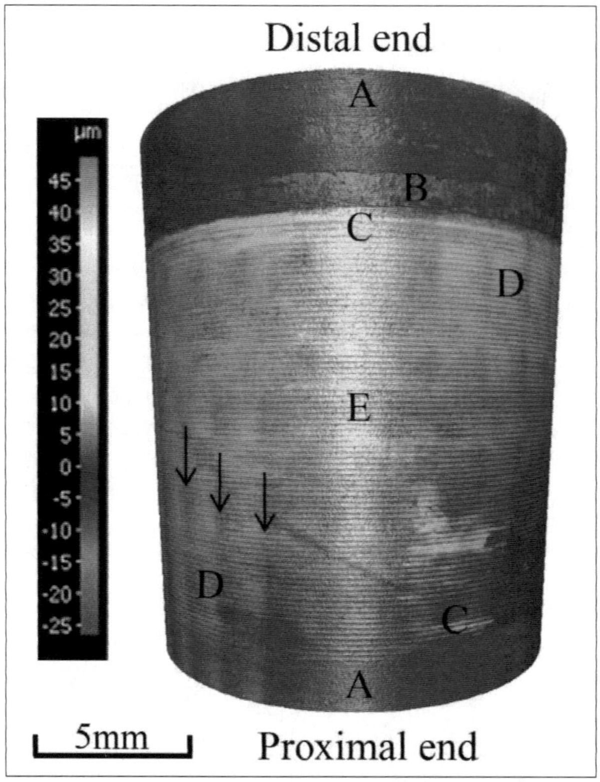

Figure 6. Deviation map of sample 5 taper replica measurement. showing A)-E) noted points of interest (see main text)

The regions labelled (D) show evidence of surface striping in the axial direction indicated by the black arrows. These stripes correspond to the spacing of the 26 longitudinal measurement regions created by the incremental steps of the FV rotary stage. The mechanism responsible for this form variation associated with stitching is not fully understood. It is thought to be related to the averaging of point height data in the overlap regions of the stitching process and is thought to be limited to the range of surface roughness heights. The effect causes a local variation in taper surface height of approximately 1.5µm amplitude with an angular wavelength of approximately 14°. It is noted that this variation is small in comparison to the real form variations.

Figure *7* and 8 show the volume loss analysis images from the Matlab™ software [12], for sample 5 measured with the RMM and FV techniques respectively. Parts (1) and (2) cover unworn region assignment with levelling and surface height thresholding respectively. Part (3) shows the Abbot-Firestone curve representation of volume loss after thresholding.

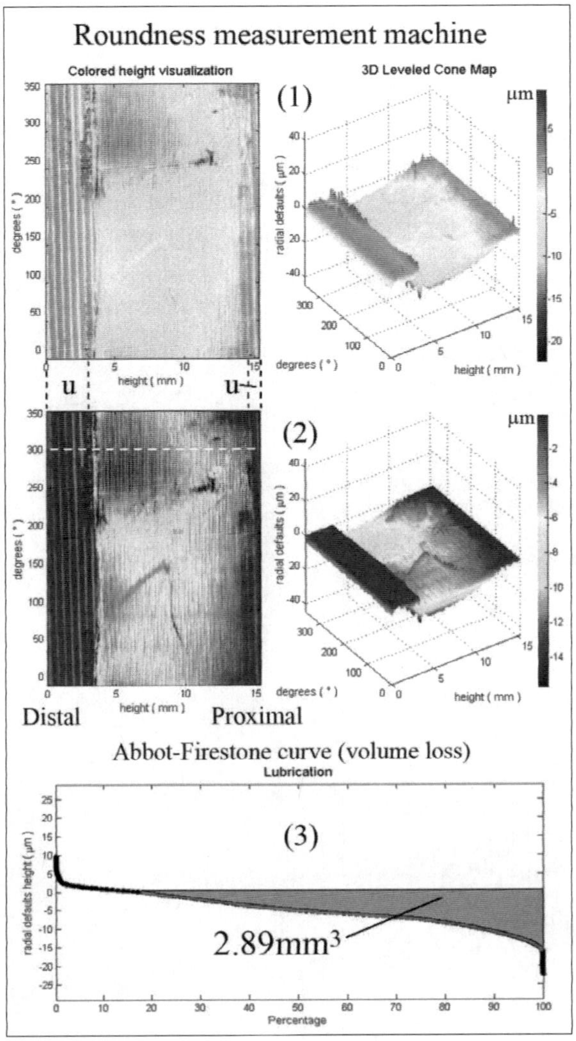

Figure 7. Volume loss analysis images for sample 5 measured by RMM. Showing the processing stages 1)-3) Unworn fitted regions are labelled (u)

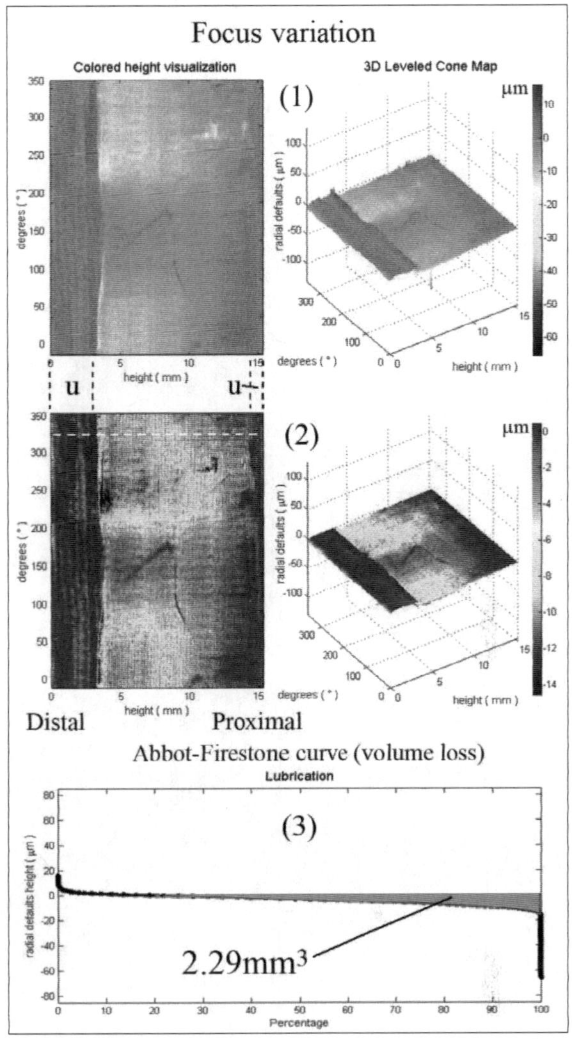

Figure 8. Volume loss analysis images for sample 5 measured by FV. Showing the processing stages 1)-3) Unworn fitted regions are labelled (u)

The two figures are not to the same scale due to outliers in the FV data, and the surface maps are offset as they have different stating points on the taper surface. The listed volume loss values correspond to those in **Table *1***, and the FV technique reports a volume loss approximately 20% lower than the RMM technique.

Figure *9* shows two profiles from the same relative location on the two surface maps from **Figure** *7* and 8 the profiles are located at the broken white lines on their respective surface maps. Despite the difference in scale the approximate broken line triangles with their heights show how the extent of wear and thus volume loss is under reported in the FV dataset. The consequences of the linear fit and levelling to zero height of the taper surface maps can be seen in **Figure *9*** (a) and (b). Importantly it should be noted that the distal unworn fitted regions of the two surfaces have different form. This difference means the zero height line to which the surfaces are thresholded is fitted at a lower relative position for the FV surface than for the RMM surface, and is largely responsible for the discrepancy in measured volume loss values.

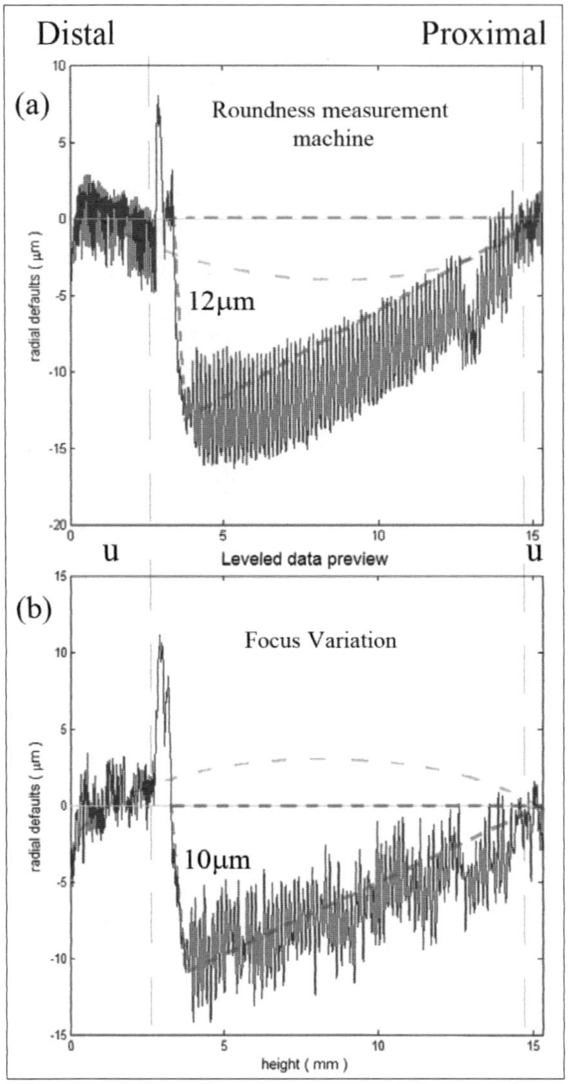

Figure 9. Levelled and linear fitted surface height map profiles for sample 5. Extracted from the locations of the white dotted lines in the respective surface height maps in Figure 7 and 8. Unworn fitted regions are labelled (u)

Figure ***9*** (a) and (b) also show broken curved lines, which, given the shape and orientation of the unworn regions might be interpreted as the likely shape of the tapers original manufactured surface. For an accurately reported surface levelling to such a polynomial fitted surface may improve the accuracy of volume loss analysis.

The type of replication discrepancy seen in **Figure** ***9*** was noted as the main replication fault during testing and eliminating it fully did not prove possible. It was to mitigate this issue that the extensions of the replica casting void as noted in **Figure** ***4*** were arranged. At the distal end, distortions in the replica caused by displacement stiction at the mounting tube junction were alleviated by moving the junction away from the measured surface by adding the extension sleeve (**Figure** ***3*** (B)). At the proximal end it was found that moving the physical end of the replica away from the measured surface reduced distortions in this region. The reason for this improvement is not clear and it is noted that in

IOP Conf. Series: Journal of Physics: Conf. Series **1183** (2019) 012004 doi:10.1088/1742-6596/1183/1/012004

the Redlux™ study that this proximal extension was not possible, nor were any such replication distortions reported. The non-curing elastomer used in the current study caused a number of difficulties with the replication process such as adhesion to and extraction of the replica. It is considered that the most significant difference between the techniques developed in the current study and that due to Redlux™ is this arrangement at the proximal end of the taper replica. It is however, not clear if this makes a significant contribution to the noted difference in accuracy.

Table *2* shows the volumetric loss data from the Redlux™ replication technique [15] for comparison to those of the current technique. The Redlux™ technique uses Microset™ fluid grade 202 blue, with quoted resolution and shrinkage values of 0.05µm and less than 0.1% respectively. This technique was validated by comparison to the gravimetric analysis of seven fabricated tapers with simulated wear designed to emulate that of typical recovered head tapers. For each taper replica 844,000 data points were acquired at half degree increments radially and approximately 14µm axial spacing, using a helical global sampling strategy. The taper validation measurements were reported over a range of volume losses of 0.46 to 4.67mm^3 and volume loss repeatability was reported for two recovered head tapers over three repeated measurements. All measured volumes were larger than their respective gravimetric values, and no correlation was noted between the magnitude of simulated wear and the difference in measured volumes. The recovered heads and manufactured tapers in the Redlux™ study were all of the same design and included no modular taper sleeve, the 12/14mm taper trunnion fitted directly into a 12/14mm taper modular head (as in **Figure** *2*).

Table 2. Volume loss data for the RedluxTM technique including repeatability data with mean, standard deviation (σ) and coefficient of variation (Cv) values [15]

Volume loss				
Sample	**Gravimetric**	**Redlux**	**Difference from Gravimetric**	
	(mm^3)		**(mm^3)**	**(%)**
1	1.72	1.73	0.01	0.58
2	2.58	2.61	0.03	1.16
3	2.22	2.32	0.1	4.50
4	0.46	0.47	0.01	2.17
5	0.72	0.86	0.14	19.44
3 (2)	4.67	4.8	0.13	2.78
4 (2)	2.78	2.82	0.04	1.44
		mean	**0.065**	**4.58**

Repeatability						
Sample	**Volume loss (mm^3)**					**Cv**
	1	2	3	**mean**	**σ**	$\sigma/mean$
a	0.55	0.85	0.57	**0.66**	**0.17**	**0.255**
b	7.47	7.15	7.24	**7.29**	**0.17**	**0.023**

Comparing the data in **Table** *1*and **Table** *2* and Redlux™ and the current technique, approximate overall ranges in volume loss difference from their respective benchmarks are noted to be 1to 19% (0.01 to 0.14 mm^3) and 1 to 24% (0.02 to 1.39mm^3) respectively. These show a similar overall spread of results, though the Redlux™ technique has better accuracy having approximate mean (absolute) values of volume difference of 5% (0.06 mm^3) compared to that of the current technique 14%

($0.4mm^3$). Mean repeatability of the two techniques is similar with standard deviations of approximately $0.17mm^3$ over the two repeated samples in each case. Again the current technique performs better relatively when coefficient of variation is used though the same caveat regarding this metric applies. Only the repeatability aspect of the Redlux™ study was carried out with retrieved heads showing real wear. Assessing the volume loss due to simulated wear is more straight forward than real in vivo wear, thus the most meaningful comparison for the current and Redlux™ techniques is the repeatability data. Therefore it is reasonable to say that the current technique performs well in comparison to the existing replication technique due to Redlux™. However, both of these replication based techniques show significant outliers in their results of the order of 20 to 30% variance, though fewer are noted in the Redlux™ data. These extreme values in both cases are reported for samples with low volume loss and thus perhaps the replication process in general may not be appropriate for measuring low volume loss. For the Redlux™ technique Cook et al [14, 15] make no comment about the possible correlation of process accuracy and scale of sample volume loss. A distinct difference is also noted in that all the Redlux™ results are reported as increases in volume difference and the opposite trend is apparent in the current technique, the reason for this and or its significance are not known.

Having noted the accuracy limitations of the current technique compared to the RMM technique it is noted that replication has the advantage of allowing the whole taper surface to be captured. Unlike a stylus technique where approximately 0.5mm of taper surface at both distal and proximal ends is not measured due to stylus access issues and run up. In certain cases this may be the difference between capturing and not capturing the crucial unworn regions on a retrieved taper. It is also noted that when used independently the current technique's performance is somewhat better given the flexibility of the Matlab™ software [12] and an experienced operator. For example the potential hogging defect of a replicated surface could be accounted for by the operator applying a polynomial level approach without the knowledge of where the form error originated. This is not to say that this approach is justified and rigorous unbiased techniques were applied in the current work.

4. Conclusions

A new volume loss analysis technique for retrieved femoral head taper surfaces has been described and its performance compared to existing techniques. Against the noted contact stylus benchmark technique mean absolute variation in volume loss is approximately 14% and repeatability is likely to be less good overall. The new technique is shown to perform well in comparison to the existing Redlux™ replication technique for retrieved tapers. Benchmarked performance is least good for tapers with small volume loss, and data may suggest this is also the case for the Redlux™ technique. Thus, replication techniques may have an inherent limitation with respect to resolving small (sub $1mm^3$) volume losses.

A characteristic type of replication defect which causes the under reporting of volume loss is noted and identified as the significant source of error in the method.

Results suggest that noise should be filtered from the point height data sets before volume loss analysis is carried out.

The script developed to unwrap the captured point cloud data to point height format has been demonstrated and allows the use of existing software which provides flexible volume loss analysis.

Further development may improve replication process control which is central to the techniques accuracy.

5. Acknowledgement

The authors gratefully acknowledge the UK's Engineering and Physical Sciences Research Council (EPSRC) funding of the EPSRC Centre for Innovative Manufacturing in Advanced Metrology (Grant Ref: EP/I033424/1).

IOP Conf. Series: Journal of Physics: Conf. Series **1183** (2019) 012004 doi:10.1088/1742-6596/1183/1/012004

Permission for the collection and analysis of the retrieved heads was granted by the National Research Ethics Service

6. References

[1] Learmonth, I.D., C. Young, and C. Rorabeck, *The operation of the century: total hip replacement.* The Lancet, 2007. **370**(9597): p. 1508-1519.

[2] No authors listed, *12th Annual report.* 2014, National Joint Registry for England and Wales: Hemel Hempstead.

[3] Kop, A.M. and E. Swarts, *Corrosion of a hip stem with a modular neck taper junction: a retrieval study of 16 cases.* The Journal of arthroplasty, 2009. **24**(7): p. 1019-1023.

[4] Gill, I., J. Webb, K. Sloan, and R. Beaver, *Corrosion at the neck-stem junction as a cause of metal ion release and pseudotumour formation.* J Bone Joint Surg Br, 2012. **94**(7): p. 895-900.

[5] Fricka, K.B., H. Ho, W.J. Peace, and C.A. Engh, *Metal-on-metal local tissue reaction is associated with corrosion of the head taper junction.* The Journal of arthroplasty, 2012. **27**(8): p. 26-31. e1.

[6] Pandit, H., S. Glyn-Jones, P. McLardy-Smith, R. Gundle, D. Whitwell, C. Gibbons, S. Ostlere, N. Athanasou, H. Gill, and D. Murray, *Pseudotumours associated with metal-on-metal hip resurfacings.* Bone & Joint Journal, 2008. **90**(7): p. 847-851.

[7] Gilbert, J.L., C.A. Buckley, and J.J. Jacobs, *In vivo corrosion of modular hip prosthesis components in mixed and similar metal combinations. The effect of crevice, stress, motion, and alloy coupling.* Journal of biomedical materials research, 1993. **27**(12): p. 1533-1544.

[8] Goldberg, J.R., J.L. Gilbert, J.J. Jacobs, T.W. Bauer, W. Paprosky, and S. Leurgans, *A multicenter retrieval study of the taper interfaces of modular hip prostheses.* Clinical orthopaedics and related research, 2002. **401**: p. 149-161.

[9] Hallab, N.J., C. Messina, A. Skipor, and J.J. Jacobs, *Differences in the fretting corrosion of metal–metal and ceramic–metal modular junctions of total hip replacements.* Journal of Orthopaedic Research, 2004. **22**(2): p. 250-259.

[10] Cooper, H.J., C.J. Della Valle, and J.J. Jacobs. *Biologic implications of taper corrosion in total hip arthroplasty.* in *Seminars in Arthroplasty*. 2012. Elsevier.

[11] Langton, D., R. Sidaginamale, J. Lord, A. Nargol, and T. Joyce, *Taper junction failure in large-diameter metal-on-metal bearings.* Bone and Joint Research, 2012. **1**(4): p. 56-63.

[12] Bills, P.J., R. Racasan, P. Tessier, and L. Blunt, *Assessing the material loss of the modular taper interface in retrieved metal-on-metal hip replacements.* Surface Topography: Metrology and Properties, 2015. **3**(2): p. 025002.

[13] Underwood, R., S. Kocagoz, R. Smith, R. Sayles, R. Siskey, S. Kurtz, and P. Cann, *A Protocol to Assess the Wear of Head/Neck Taper Junctions in Large Head Metal-on-Metal (LHMoM) Hips*, in *Metal-On-Metal Total Hip Replacement Devices*. 2013, ASTM International.

[14] Cook, R.B., B.J. Bolland, J.A. Wharton, S. Tilley, J.M. Latham, and R.J. Wood, *Pseudotumour formation due to tribocorrosion at the taper interface of large diameter metal on polymer modular total hip replacements.* The Journal of arthroplasty, 2013. **28**(8): p. 1430-1436.

[15] Cook, R., C. Maul, and A. Strickland, *Validation of an Optical Coordinate Measuring Machine for the Measurement of Wear at the Taper Interface in Total Hip Replacement*, in *Modularity and Tapers in Total Joint Replacement Devices*. 2015, ASTM International.

[16] BS EN ISO_25178-606:2015, *Surface texture: Areal Part 606: Nominal characteristics of non-contact (focus variation) instruments*, in *Geometrical product specification (GPS)* 2013, British Standards Institute.

[17] Nayar, S.K. *Shape from focus system*. 1992. IEEE.

[18] Danzl, R., F. Helmli, and S. Scherer, *Focus Variation--a Robust Technology for High Resolution Optical 3D Surface Metrology.* Strojniski Vestnik/Journal of Mechanical Engineering, 2011. **57**(3).

[19] Walton, K., L. Blunt, L. Fleming, M. Goodhand, and H. Lung, *Areal parametric characterisation of ex-service compressor blade leading edges.* Wear, 2014. **321**: p. 79-86.
[20] BS EN ISO_10360-8:2013, *Acceptance and reverification tests for coordinate measuring systems (CMS) Part 8: CMMs with optical distance sensors*, in *Geometrical produc specifications (GPS)*. 2013, British Standards Institute.
[21] BS EN ISO_10360-3:2001, *Acceptance and reverification tests for coordinate measuring machines (CMM)*, in *Geometrical produc specifications (GPS)*. 2001, British Standards Institute.
[22] Sun, W. and J.D. Claverley, *Verification of an optical micro-CMM using the focus variation technique: Aspects of probing errors.* CIRP Annals-Manufacturing Technology, 2015. **64**(1): p. 511-514.
[23] Moroni, G., W.P. Syam, and S. Petrò, *On Performance Verification of Simultanous 4-axis 3D geometric Optical Instrument.*
[24] Abbot, E. and F. Firestone, *Specifying surface quality.* Mech. Eng, 1933. **55**: p. 569.
[25] BS EN ISO_13565-2:1997, Surface texture: Profile method. Surfaces having stratified functional properties. Height characterization using the linear material ration curve, in Geometrical product specifications (GPS). 1997, British Standards Institute.
[26] BS EN ISO_25178-3:2012, *Surface texture: Areal Part 3: Specification operators*, in *Geometrical product specifications (GPS)* 2012, British Standards Institute.

IOP Conf. Series: Journal of Physics: Conf. Series **1183** (2019) 012005 doi:10.1088/1742-6596/1183/1/012005

Surface Topography of Nitrided Steel Surfaces

Johan Berglund[1,2] and Eva Troell[1]
[1]Department of Manufacturing, Swerea IVF, PO Box 104, SE-43122 Mölndal, Sweden
[2]Department of Industrial and Materials Science, Chalmers University of Technology, SE-41296 Gothenburg, Sweden

E-mail: johan.berglund@swerea.se

Abstract. The effect of nitrocarburizing were investigated for two different steels, 42CrMo4 and 25CrMo4, with objective to evaluate the influence of initial surface topography on the resulting nitrocarburized surface with regard to surface topography and thickness of the compound layer. It was found that the nitrocarburizing process has an impact on the surface topography. The process creates a short-wave isotropic structure on the original surface and this is particularly evident for the smoother original surfaces. No significant effect on the compound layer thickness depending on the surface topography before heat treatment could be observed.

1. Introduction

Currently, case hardening is the most common heat treatment method for larger volumes of high strength steel components. Nitriding and nitrocarburizing are some of the more interesting alternatives [1], [2]. These processes allow in several ways reduced environmental loads, both at production and during the life time of the product in use. In many cases components with low friction and good wear resistance are desired, e.g hydraulic components, pistons, gear wheels and tools [3], [4].

Compared to case hardening additional production benefits are provided such as less distortion and reduced energy consumption [3]. During the service life of the product commonly obtained benefits are improved component performance due to good surface properties (friction and wear) [5], [6], improved fatigue strength [3], [7] and increased corrosion resistance [8], [9]. These properties are also known to be affected by the surface topography. Low friction powertrain-components are needed in order to meet requirements for lower fuel consumption in the automotive industry. For these kinds of applications nitriding processes, combined with a suitable steel, are of great interest [3].

The surface topography of the heat treated component will be influenced by the nitriding process [10]. It is additionally hypothesised that the result will depend on the topography of the original surface, before treatment. However, this effect is not well documented in the literature. In many studies the samples are prepared to have the same initial surface characteristics [10]–[12].

The objective of the present study was to evaluate the influence of initial surface topography on the resulting nitrocarburized surface with regard to surface topography and thickness of the compound layer after nitrocarburizing.

2. Materials and methods

Two different steel materials were used in the present investigation, 42CrMo4 and 25CrMo4. The steels were atmospheric nitrocarburized for 150 min in 580°C. With the 42CrMo4 steel 4 samples were created with different surface finish by polishing to different levels. The 25CrMo4 steel was used in a real hydraulic motor component from which 5 samples were taken with surfaces created by different machining processes, see Table 1. The compound layer has been analysed using Light Optical Microscopy (LOM).

Surface topography of the samples was measured and analysed before and after heat treatment. The 42CrMo4 samples were relocated so that the same area could be analysed before and after heat treatment. The relocation was performed by means of indentations on the surface that were visible. The indentations were used to shift and rotate the surface so that the same regions were measured before and after treatment. The 25CrMo4 samples were not relocated due to geometry dependent practical reasons. The topography measurements were made using an Olympus LEXT OLS3100 confocal instrument [13].Stitching (3x3) and analysis were made with the software MountainsMap from Digital Surf. Measurement size and resolution were chosen depending on the surface texture. Form removal with 2'nd order polynomial and noise reduction was applied to the measured surfaces before analysis to provide S-F surfaces. The data sets were also cropped to create pairs of the same size, before and after heat treatment. See Table 1 for details on the analysed data sets.

Selected surface roughness parameters were calculated according to ISO 25178-2 [14]. The parameters Sa, Sdr and Str were selected since they represent different characteristics of the surface that were thought to be relevant to study. Sa, arithmetical mean height of the surface, is a commonly used parameter for areal characterisation of surface roughness. In many ways it is a areal equivalent to the traditionally used profile Ra [15]. However, the use of Ra also implies the use of certain filtering, which is not the case for Sa [16]. Sdr, developed interfacial area ratio, is used as a measure of the surface complexity, especially in comparisons between several stages of processing on a surface. It can be helpful with discriminating between surfaces with similar Sa but with different textures within the shorter wavelengths since longer wavelengths on a surface tend to have higher amplitudes than shorter wavelengths and will often therefore dominate the Sa value [17], [18]. Str, texture aspect ratio, characterises the isotropy of the surface. The Str parameter is unit-less and its values is between 0 and 1. If Str is close to 1 the surface is isotropic, which means that it has the same properties regardless of the direction. If Str is close to 0 the surface is anisotropic, which means that it has a dominant texture direction [16].

Table 1. Samples materials, size of analysed area and lateral resolution for topography measurements.

Sample	Steel	Analysed area (μm)	Lateral resolution (μm)	Objective lens
80 grit paper	42CrMo4	1560 x 1560	0.625	20x
320 grit paper	42CrMo4	1250 x 1250	0.625	20x
800 grit paper	42CrMo4	600 x 600	0.250	50x
8000 grit paper	42CrMo4	625 x 625	0.250	50x
Ground	25CrMo4	650 x 650	0.250	50x
Rough turned	25CrMo4	1310 x1310	0.625	20x
Finish turned	25CrMo4	1370 x 1370	0.625	20x
Finish milled	25CrMo4	625 x 625	0.250	50x
Reamed	25CrMo4	1620 x 1620	0.625	50x

3. Results: Surface topography

3.1. Steel 42CrMo4 prepared with 80 grit paper
Before and after heat treatment, see **Figure 1**. Observations:

- Lower Sa, smoother surface
- Slight increase in Sdr, more presence of a shorter wavelength texture
- Str is similar, the bigger scratches dominates and are still apparent after heat treatment

IOP Conf. Series: Journal of Physics: Conf. Series **1183** (2019) 012005 doi:10.1088/1742-6596/1183/1/012005

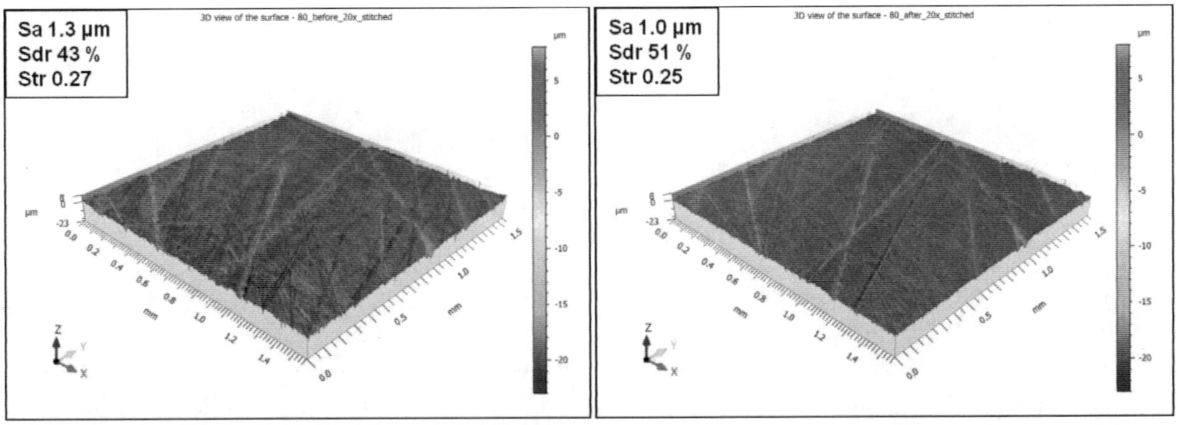

Figure 1. Topography of surface '80 grit paper' before (left) and after (right) nitrocarburizing, relocated. Steel 42CrMo4.

3.2. Steel 42CrMo4 prepared with 320 grit paper

Before and after heat treatment, see **Figure 2**. Observations:

- Sa is similar
- Increase in Sdr, more presence of a shorter wavelength texture
- Clear anisotropic texture with polishing scratches before, low Str. After nitrocarburizing the surface is different and isotropic, high Str

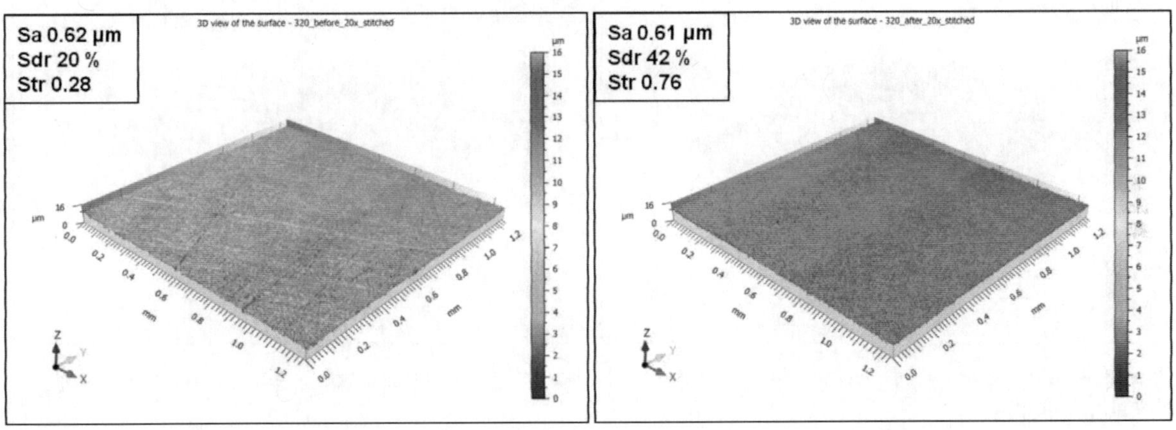

Figure 2. Topography of surface '320 grit paper' before (left) and after (right) nitrocarburizing, relocated. Steel 42CrMo4.

3.3. Steel 42CrMo4 prepared with 800 grit paper

Before and after heat treatment, see **Figure 3**. Observations:

- Before the surface is smooth with polishing scratches. Low Sa, Sdr and Str
- After, nothing of the previous texture can be seen, instead there is a much rougher isotropic texture. Higher Sa, Sdr and Str

IOP Conf. Series: Journal of Physics: Conf. Series **1183** (2019) 012005 doi:10.1088/1742-6596/1183/1/012005

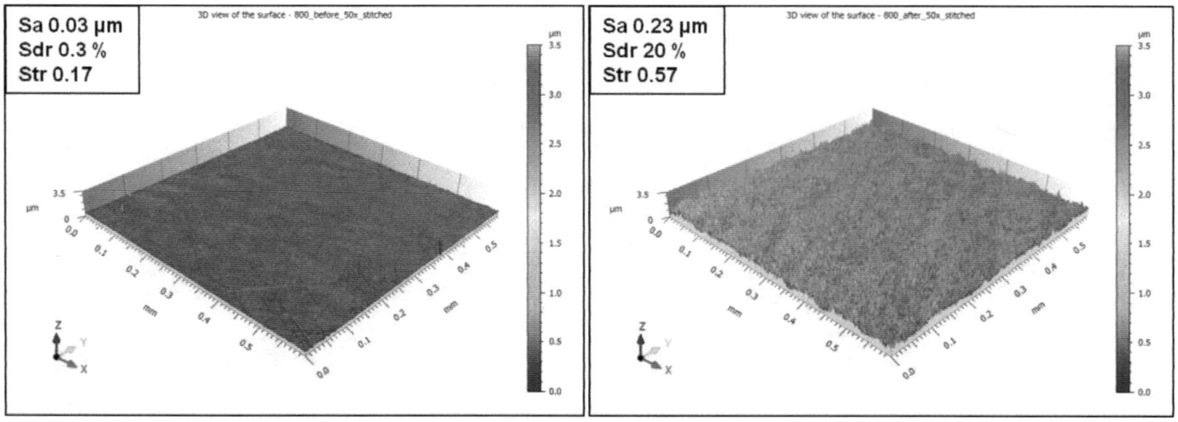

Figure 3. Topography of surface '800 grit paper' before (left) and after (right) nitrocarburizing, relocated. Steel 42CrMo4.

3.4. Steel 42CrMo4 prepared with 4000 grit paper

Before and after heat treatment, see **Figure 4**. Observations:

- Before the surface is smooth with some polishing scratches. Low Sa, Sdr and medium Str
- After, nothing of the previous texture can be seen, instead there is a much rougher and more isotropic texture. Higher Sa, Sdr and Str

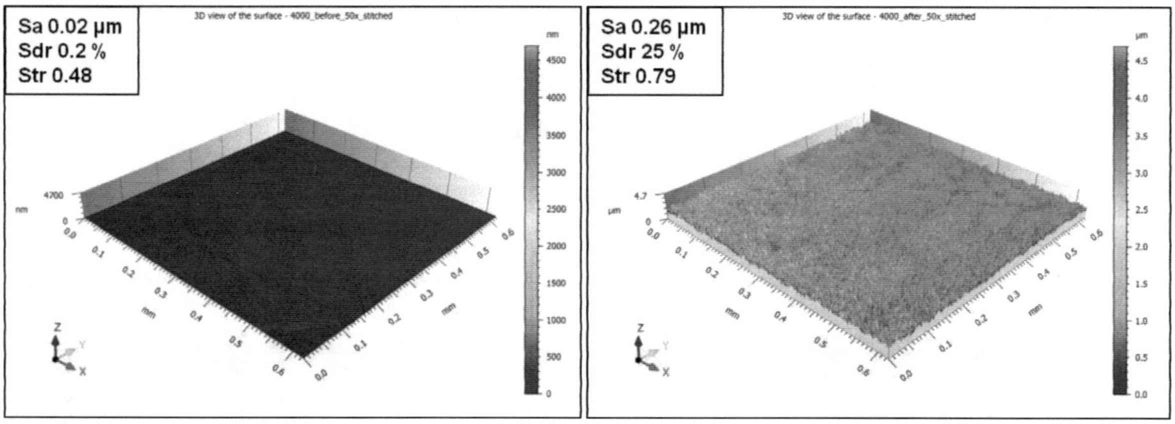

Figure 4. Topography of surface '4000 grit paper' before (left) and after (right) nitrocarburizing, relocated. Steel 42CrMo4.

3.5. Steel 25CrMo4 ground

Before and after heat treatment, see **Figure 5**. Observations:

- Similar Sa
- Slight increase in Sdr
- No big difference before and after heat treatment

IOP Conf. Series: Journal of Physics: Conf. Series **1183** (2019) 012005 doi:10.1088/1742-6596/1183/1/012005

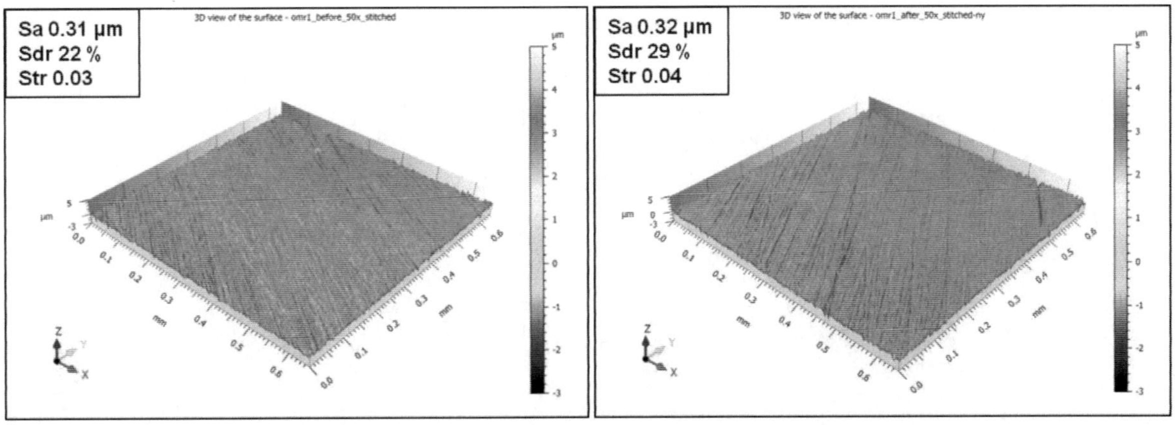

Figure 5. Topography of surface 'Ground' before (left) and after (right) nitrocarburizing, not relocated. Steel 25CrMo4.

3.6. Steel 25CrMo4 rough turned

Before and after heat treatment, see **Figure 6**. Observations:

- Increase in Sdr, more presence of a shorter wavelength texture
- Sa and Str are dominated by machining signature

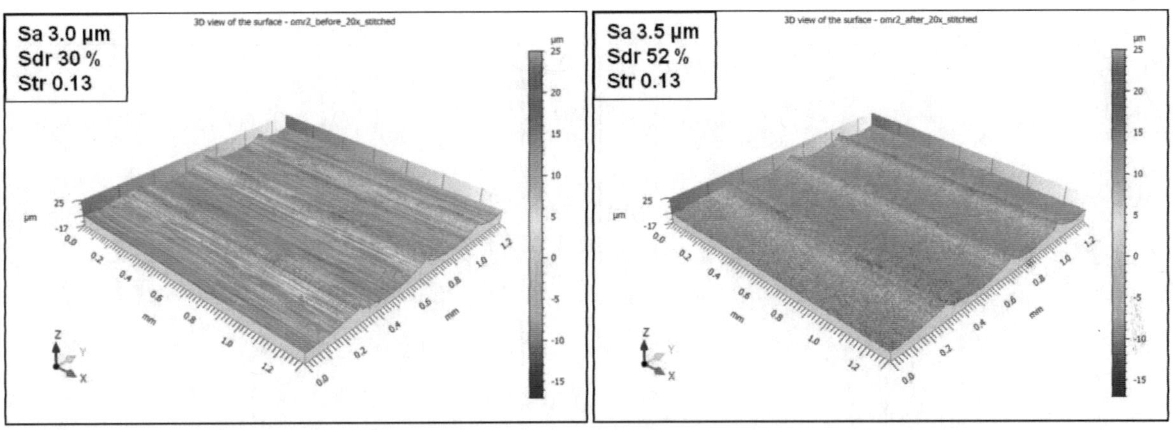

Figure 6. Topography of surface 'Rough turned' before (left) and after (right) nitrocarburizing, not relocated. Steel 25CrMo4.

In **Figure 7** a sub-region of 'Rough turned' is presented. An area between two turning feed ridges have been selected for further study. An additional form removal step has been performed (3'rd order polynomial). Here it can be observed that Sdr and Str has increased and Sa has decreased.

IOP Conf. Series: Journal of Physics: Conf. Series **1183** (2019) 012005 doi:10.1088/1742-6596/1183/1/012005

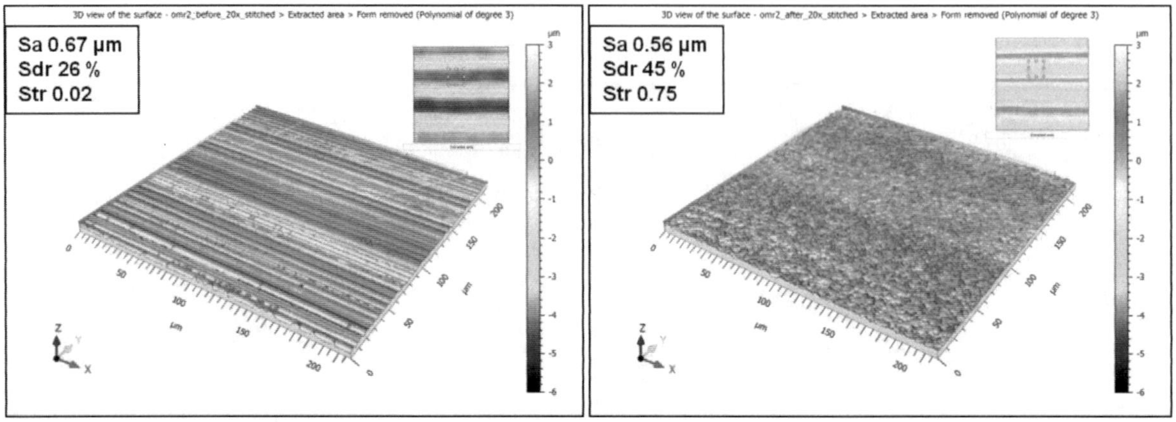

Figure 7. Detail of topography of surface 'Rough turned' before (left) and after (right) nitrocarburizing, not relocated. Additional cropping and form removal (3'rd order polynomial) has been applied. Steel 25CrMo4.

3.7. Steel 25CrMo4 finish turned

Before and after heat treatment, see **Figure 8**. Observations:

- Similar before and after heat treatment
- Sdr has decreased

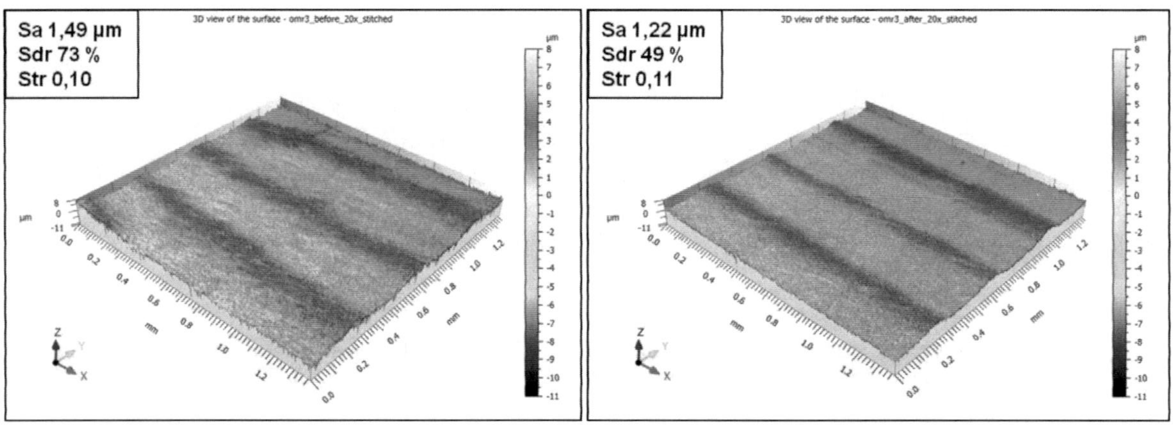

Figure 8. Topography of surface 'Finish turned' before (left) and after (right) nitrocarburizing, not relocated. Steel 25CrMo4.

3.8. Steel 25CrMo4 finish milled

Before and after heat treatment, see **Figure 9**. Observations:

- The surface is rougher after heat treatment, Sa has increased
- Sdr and Str increases
- The anisotropic texture from milling is mixed with an isotropic short wave texture

IOP Conf. Series: Journal of Physics: Conf. Series **1183** (2019) 012005 doi:10.1088/1742-6596/1183/1/012005

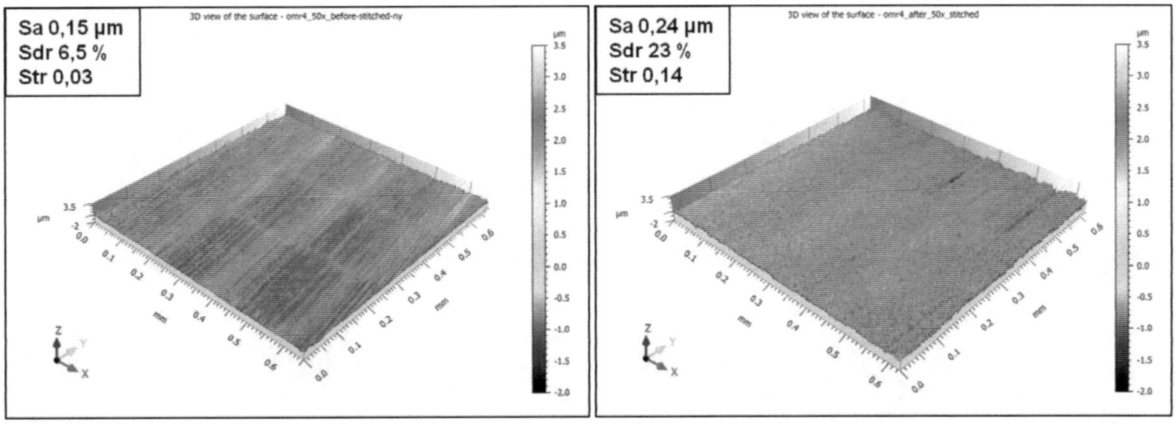

Figure 9. Topography of surface 'Finish milled before (left) and after (right) nitrocarburizing, not relocated. Steel 25CrMo4.

3.9. Steel 25CrMo4 reamed

Before and after heat treatment, see **Figure 10**. Observations:

- The surface is rougher after heat treatment, Sa and Sdr increases.
- Str is low before and after. The surface is dominated by large scratches which are not so much affected by the nitrocarburizing process.

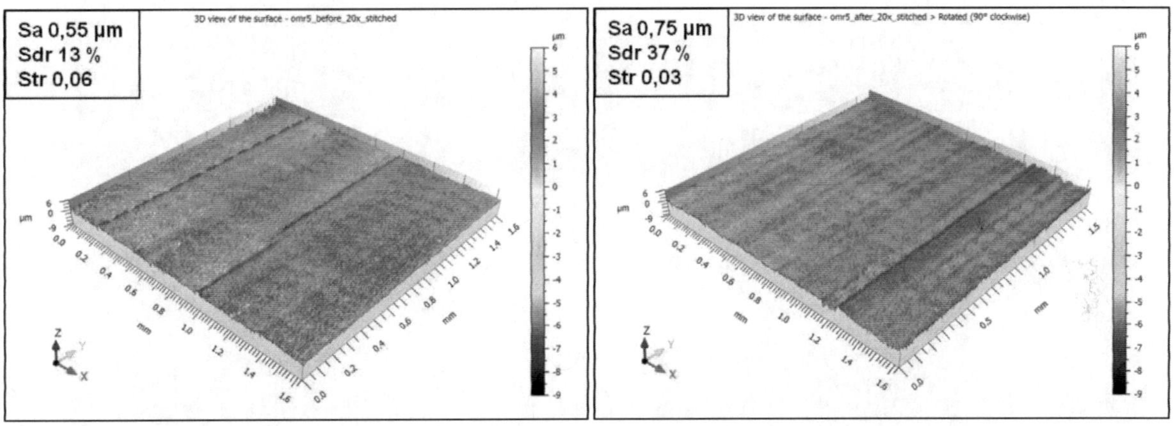

Figure 10. Topography of surface 'Reamed' before (left) and after (right) nitrocarburizing, not relocated. Steel 25CrMo4.

4. Results: Compound layer

For all the surfaces on the component in steel 25CrMo4 the compound layer thickness was 11-12 µm, see **Figure 11** a-e. In this case no effect on the compound layer thickness depending on the surface topography before heat treatment could be noticed. However the final surface topography is influenced by the machining process, e.g. when comparing the ground and rough turned surfaces.

IOP Conf. Series: Journal of Physics: Conf. Series **1183** (2019) 012005 doi:10.1088/1742-6596/1183/1/012005

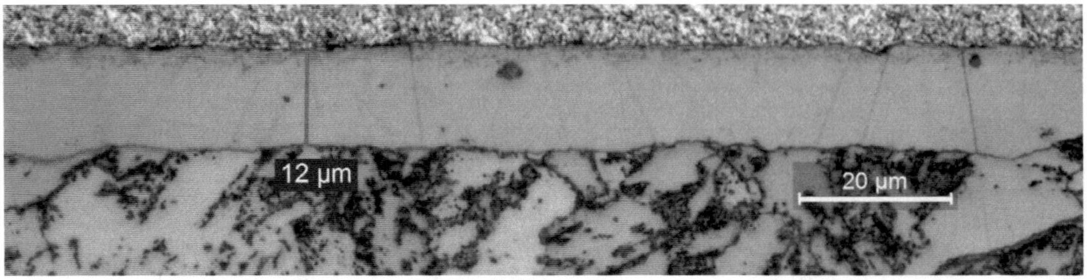

a) Ground

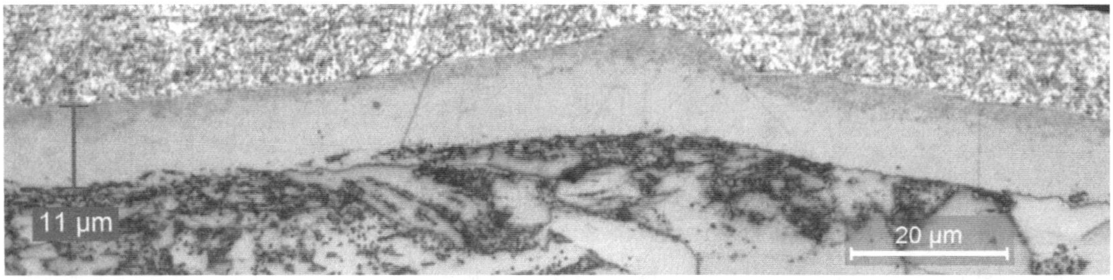

b) Rough turned

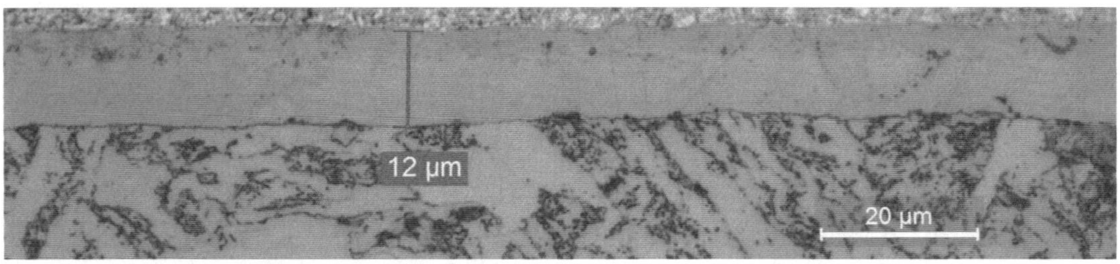

c) Finish turned

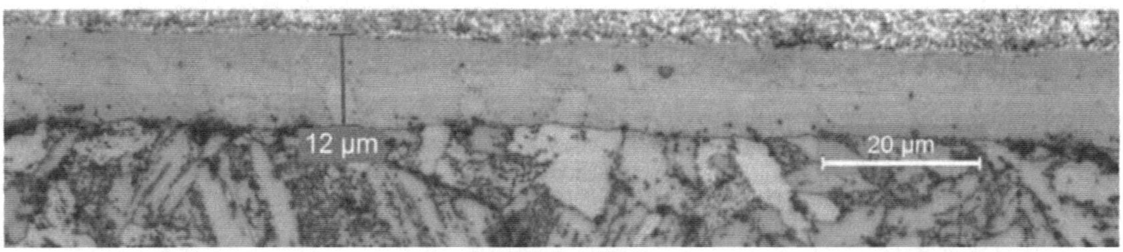

d) Finish milled

e) Reamed

Figure 11 a-e. Compound layer on surfaces of steel 25CrMo4 for surfaces created with different machining processes after nitrocarburizing, 580°C 150 min.

IOP Conf. Series: Journal of Physics: Conf. Series **1183** (2019) 012005 doi:10.1088/1742-6596/1183/1/012005

5. Analysis and discussion

Since the 42CrMo4 surfaces were relocated it was possible to do a precise analysis of the differences before and after nitrocarburizing regarding the surface topography. It was clear that the nitrocarburizing process has an impact on the surface topography. The process creates a short-wave isotropic structure on the original surface and this is particularly evident for the smoother original surfaces (Sa < 0.2 μm). The 25CrMo4 surfaces were not relocated so the exact same areas could not be analysed before and after nitrocarburizing. This makes drawing conclusions more difficult. However, similar tendencies could be observed here as with the 42CrMo4 samples.

For one of the surfaces, Steel 25CrMo4 finish turned, Sdf showed a decrease instead of an increase which was the case for the other surfaces. Also Sa showed a decrease. The reason for the different results for this particular surface is not clear. One uncertainty is the lack of relocation which means that it is not certain that the measurements before and after nitrocarburizing were made in the same location.

No significant effect on the compound layer thickness depending on the surface topography before heat treatment could be noticed. It is possibly due to that the nitrocarburizing time was long enough to enable a stable process and thickness of the layer. The processing parameters were selected to achieve good compound layers. A shorter nitrocarburizing time may have produced samples with differentiated surface properties, both regarding thickness of the compound layer along the surface and the topography. Also, it is possible that initial sample surfaces with more short wavelength texture would react differently to the nitrocarburizing process.

6. Conclusions

- The nitriding processes has an impact on the surface topography, thus the original surface will be affected by the nitrocarburizing.
- A short-wavelength isotropic texture is created on the original surface. Particularly evident for smoother original surfaces (Sdr and Str increases)
- If the original surface is very smooth the surface will be rougher when the short-wavelength texture is created (Sa increases)

7. Acknowledgement

The authors wish to acknowledge Vinnova, a Swedish government agency working under the Ministry of Enterprise and Innovation, for the financial support through the FFI program of the project “Surfaces with optimal friction and fatigue properties of nitrided/nitrocarburized components. Production, performance and sustainability aspects - SurfNit”, DNr 2013-05600, during which all the research work presented in the current paper have been performed.

8. References

[1] T. Holm, P. Olsson, and E. Troell, *Steel and Its Heat Treatment: A Handbook*. SWEREA IVF, 2012.

[2] D. Pye, ‘Nitriding and Nitrocarburizing’, in *Encyclopedia of Tribology*, Springer, Boston, MA, 2013, pp. 2421–2428.

[3] E. Troell, J. Senaneuch, N. Hawsho, and H. Jesperson, ‘Fatigue and wear properties of nitrided steels’, Internal report VBC-R-2013-6, 2013.

[4] J. Vetter, ‘Surface Treatments for Automotive Applications’, in *Coating Technology for Vehicle Applications*, Springer, Cham, 2015, pp. 91–132.

[5] H. Kato, T. S. Eyre, and B. Ralph, ‘Sliding wear characteristics of nitrided steels’, *Surf. Eng.*, vol. 10, no. 1, pp. 65–74, Jan. 1994.

[6] A. Molinari, G. Straffelini, M. Pellizzari, and M. Pirovano, ‘Wear behaviour of diffusion and compound layers in nitrided steels’, *Surf. Eng.*, vol. 14, no. 6, pp. 489–496, Jan. 1998.

[7] T. Bell and D. H. Thomas, ‘Cyclic stressing of gas nitrocarburized low carbon steel’, *Metall. Trans. A*, vol. 10, no. 1, pp. 79–84, Jan. 1979.

IOP Conf. Series: Journal of Physics: Conf. Series **1183** (2019) 012005 doi:10.1088/1742-6596/1183/1/012005

[8] M. Emami, H. M. Ghasemi, and J. Rassizadehghani, 'High temperature tribological behaviour of 31CrMoV9 gas nitrided steel', *Surf. Eng.*, vol. 26, no. 3, pp. 168–172, Apr. 2010.
[9] E. Boztepe, A. C. Alves, E. Ariza, L. A. Rocha, N. Cansever, and F. Toptan, 'A comparative investigation of the corrosion and tribocorrosion behaviour of nitrocarburized, gas nitrided, fluidized-bed nitrided, and plasma nitrided plastic mould steel', *Surf. Coat. Technol.*, vol. 334, pp. 116–123, Jan. 2018.
[10] R. Sola, G. Poli, P. Veronesi, and R. Giovanardi, 'Effects of Surface Morphology on the Wear and Corrosion Resistance of Post-Treated Nitrided and Nitrocarburized 42CrMo4 Steel', *Metall. Mater. Trans. A*, vol. 45, no. 6, pp. 2827–2833, Jun. 2014.
[11] P. Cavaliere, A. Perrone, and A. Silvello, 'Multi-objective optimization of steel nitriding', *Eng. Sci. Technol. Int. J.*, vol. 19, no. 1, pp. 292–312, Mar. 2016.
[12] B. Wang, S. Sun, M. Guo, G. Jin, Z. Zhou, and W. Fu, 'Study on pressurized gas nitriding characteristics for steel 38CrMoAlA', *Surf. Coat. Technol.*, vol. 279, pp. 60–64, Oct. 2015.
[13] R. Artigas, 'Imaging Confocal Microscopy', in *Optical Measurement of Surface Topography*, Springer, Berlin, Heidelberg, 2011, pp. 237–286.
[14] the International Organization for Standardization, 'ISO 25178-2:2012 Geometrical product specifications (GPS) — Surface texture: Areal — Part 2: Terms, definitions and surface texture parameters'. 2012.
[15] International Organization for Standardization, 'ISO 4287:1997 - Geometrical Product Specifications (GPS) -- Surface texture: Profile method -- Terms, definitions and surface texture parameters'. International Organization for Standardization, 1997.
[16] F. Blateyron, 'The Areal Field Parameters', in *Characterisation of Areal Surface Texture*, Springer, Berlin, Heidelberg, 2013, pp. 15–43.
[17] Tom R. Thomas, *Rough Surfaces, 2nd Edition*. World Scientific, 1998.
[18] J. Berglund, 'Characterisation of Functional Pressing Die Surfaces', Doctoral Thesis, Chalmers University of Technology, Göteborg, 2011.

IOP Conf. Series: Journal of Physics: Conf. Series **1183** (2019) 012006 doi:10.1088/1742-6596/1183/1/012006

Characterization of subsurface deformation of turned brasses: lead brass ($CuZn_{39}Pb_3$) and lead free brass ($CuZn_{21}Si_3P$)

Vijeth V Reddy[1], Pui Lam Tam[2], Amogh Vedantha Krishna[1] and B-G Rosén[1]

[1]Functional Surfaces Research Group, Halmstad University, Halmstad, Sweden

[2]Department of Industrial and Materials Science, Chalmers University of Technology, Gothenburg, Sweden

Email: vijred@hh.se

Abstract. The adverse effects of lead on human health and the recycling problems of copper alloys with lead content has led to the increase in concern for reducing/eliminating the use of lead in brass and other copper alloys. The real challenge prevails in sustainable manufacturing of lead free brass and to maintain as well as control the surface integrity when lead is substituted in the brass with silicon. This article is part of the study that focuses on characterizing the surface integrity of machined lead brass and lead free brass. This paper deals with the study of surface layer characteristics of brass samples, investigating the subsurface deformation influenced by the turning operation. It is important to study the plastic deformation under the machined surface which directly affects the performance of the material on further machining or application. In this study, metallographic investigations are performed on turned samples of lead brass and lead free brass and include comparisons between the study samples based on the plastic deformations quantified by grain orientations and microhardness measurements. The study performed successfully characterizes the subsurface deformation and the results suggest slightly higher subsurface deformation in turned lead brass as compared to lead free brass.

1. Introduction

The surface quality greatly influences the functional behaviour of a component and manufacturing efficiency and hence the study of surface integrity is an integral part in sustainable manufacturing. The addition of lead in brass greatly reduces the cutting forces and decrease in attained tool wear as compared to the lead free alternative [1]. But considering the dangerous consequences of lead contamination, manufacturing sector is increasingly focusing on identifying/ developing a contemporary manufacturing technique to increase the functionality, efficiency and to economically produce the lead free brass components. The initial investigations on surface topography of turned brass samples revealed the presence of higher surface asperities in lead brass compared to lead free brass [2].

1.1. Subsurface Deformation

Large temperature gradient during machining is attributed to the shear interaction between the tool and the workpiece [3]. This induces localized heating and plastic deformations in the subsurface region which may result in mechanical, thermal and metallurgical transformations [3, 4]. The surface properties of such strain hardened zones are central to the formation of wear debris [5] and accelerate corrosion [6]. These effects in mechanical and tribological behaviour of the surface influence the functionality and efficiency of the component. The objective is to identify and evaluate the effects of machining on surface integrity of lead brass and lead free brass. In this study, turned samples of lead- and lead free brass are investigated.

Several studies [7-10] suggested machining induced strain hardening which led to higher hardness values at the near surface region below the machined surface and gradually decrease to the bulk material hardness. In this study, the plastic deformation and subsequently induced strain hardening in the subsurface are determined by the microhardness measurements. The microhardness values of lead brass and lead free brass are evaluated as a function of depth from the surface edge. Using the microhardness measurements, the depth of deformation zone for the study samples is analysed. Depth of deformation zone refers to the depth at which the hardness values attain the bulk material hardness.

Investigations [7,10,11] on the subsurface region of machined study samples suggest that the mechanical and thermal loadings during machining induce changes in microstructural attributes like grain size, orientation and dislocation density. Metallurgical alterations like elongation of grains and grain boundaries are predominantly due to plastic deformation caused by the shear forces between the tool and the workpiece during machining.

The study material's composition is listed in table 1.

Table 1: Chemical composition according to nominal standards (wt. %), [12][13]

Material	Cu	Zn	Pb	Si	P
$CuZn_{39}Pb_3$ (Lead brass)	57.3	Balance	3.3	-	-
$CuZn_{21}Si_3P$ (Lead free brass)	76	Balance	-	3	0.05

It is well known that the material's composition is critical to evaluate the effect of machining on surface integrity. Due to the compositional difference, hardness values differ between study samples with lead free brass harder than its counterpart [14]. Hence the study samples are compared by their hardness gradient with the function of depth from surface edge. The metallurgical alterations in the subsurface region are analysed using backscattered images from Scanning Electron Microscope (SEM). The grain contrast imaging using backscattered images is discernible, fast and straight-forward compared to other techniques [15].

2. Materials and Methods

To identify and quantify the subsurface deformation, metallographic investigations are performed on the turned lead and lead free brass samples.

2.1. Study samples

The study samples include cylindrical lead- and lead free brass samples subjected to turning operation using uncoated H10F cemented carbide tool primarily containing fine grain tungsten carbide with tool geometry DNGA150708F defined by ISO 1832:2012 [16] with edge radius of approx. 10 ± 5μm. The samples are machined at cutting speed of 400m/min, 0.8mm depth of cut and 0.2 mm/rev feed rate.

2.2. Metallographic sample preparation

The samples are sectioned by wire EDM perpendicular to both the cutting direction and feed direction. The sectioned samples are mounted, as shown in figure 1, using Buehler's Konductomet™, phenolic thermoset compression mounting compound to retain the surface edge during subsequent metallographic steps. The mounted samples are ground and polished using decreasing grit size disks and low napped cloths. The cross sectional plane is etched with solution- mixture of ferric chloride (FeCl3), conc. hydrochloric acid (HCl) and distilled water in the proportion of 1:5:20. The etched samples are observed under Scanning Electron Microscope (SEM) to identify the deformation in the subsurface region.

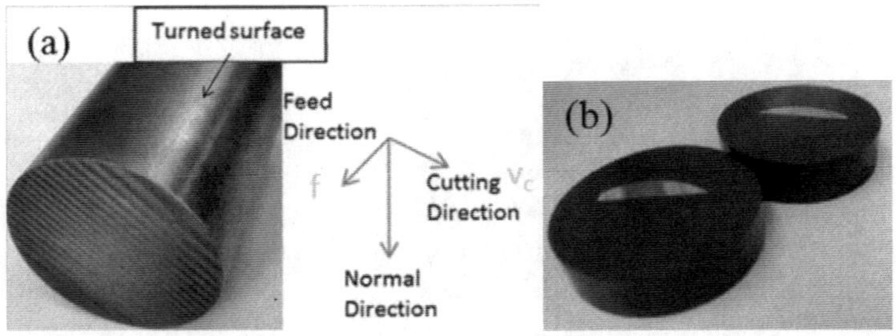

Figure 1. (a) Turned brass sample (b) Mounted section of brass sample

2.3. Vickers microhardness test

In this study, the Vickers microhardness, as specified by ISO 6507-1:2005 [17], is measured on the polished surface using the Shimadzu microhardness tester equipped with a diamond pyramid indenter. Indent load of 50g for 12 seconds is applied on the test samples in the region of interest. The measurements are taken at 10μm in depth and 90μm in lateral directions, as shown in the figure 2, with the first indent at 20 μm from the edge due to the limitation on indenter size and indent shape acquired. Vickers microhardness test includes measuring the diagonals of the indent D1 (μm), D2 (μm) and equating in (1)

$$HV = \frac{1.8544 * F}{[(D1) * (D2)]} \tag{1}$$

Where, HV is the Vickers hardness number, F is the load applied.

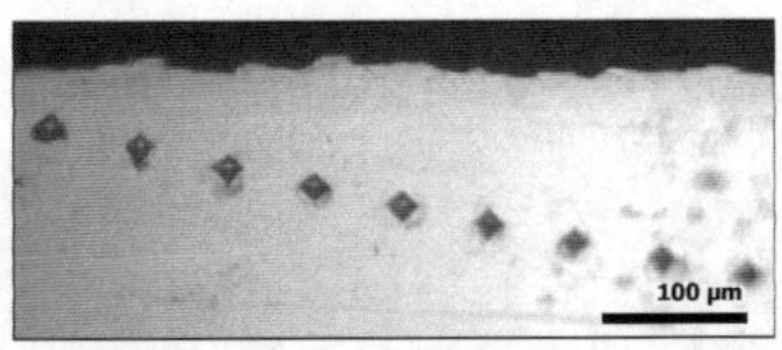

Figure 2. Vickers microhardness indentations

2.4. Grain analysis

The metallurgical alterations are analyzed using the backscattered (BSE) images of etched brass samples captured at 500X magnification with an investigation area of 250μm x 150μm. BSE detectors capture high contrast images which distinguish grains and grain boundaries. The orientation of the grains in the subsurface region is measured using Digital Surf's Mountains Map®. The image processing included binary segmentation, application of morphological filters and exclusion of incomplete grains at the borders for significant grain analysis, as shown in figure 3a. The grain orientations are the angle, as illustrated in figure 3b, by the major axis of individual grains with the horizontal plane and the average orientations are plotted as a function of depth from the surface edge.

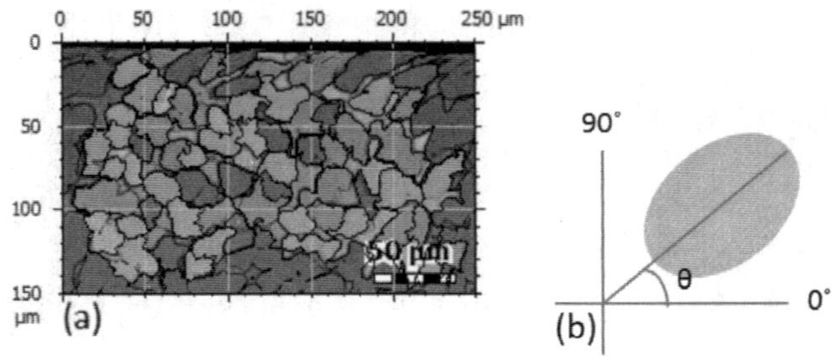

Figure 3. (a) Image processing in Mountains Map® (b) Illustration of grain orientation (θ)

3. Results and Discussions

The backscattered electron images of lead- and lead free brass are sectioned perpendicular to the feed direction, f and cutting direction, v_c to identify and quantify the subsurface deformation.

3.1. Qualitative analysis

The backscattered electron images of the subsurface region sectioned perpendicular to feed direction exhibit less- or no significant deformation in both lead- and lead free brass, as shown in figure 4, though fine grains were observed in some regions. But in the subsurface region perpendicular to the cutting direction, significant deformation exists because of higher shear in cutting direction compared to feed direction.

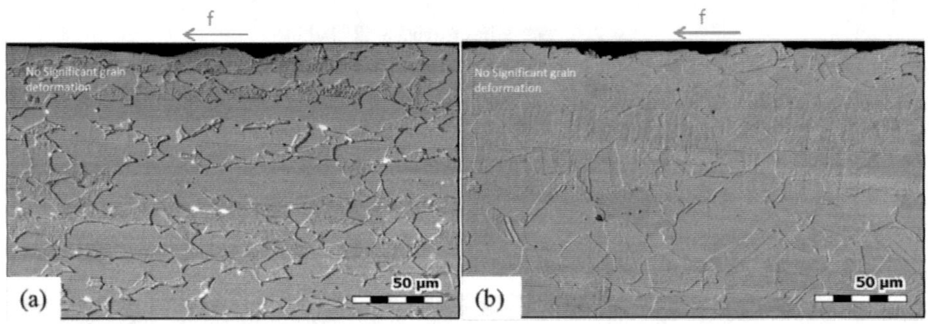

Figure 4. Etched subsurface region of (a) lead brass and (b) lead free brass samples sectioned perpendicular to the feed direction, f

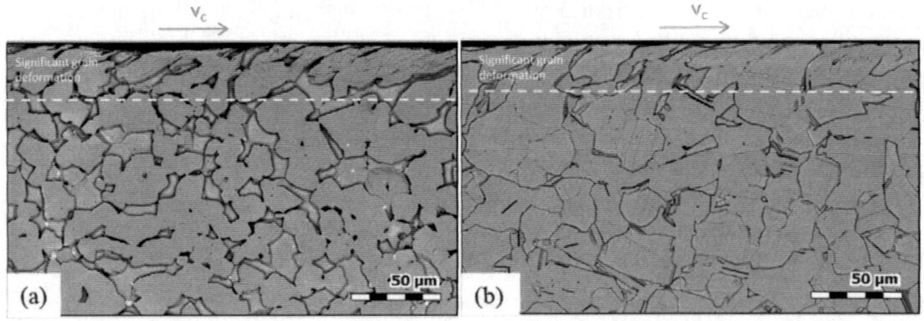

Figure 5. Etched subsurface region of (a) lead brass and (b) lead free brass samples sectioned perpendicular to the cutting direction, v_c. The delineated line approximately separates the deformed grain with the bulk material

IOP Conf. Series: Journal of Physics: Conf. Series **1183** (2019) 012006 doi:10.1088/1742-6596/1183/1/012006

In general, the grains are distributed randomly but in the region just beneath the machined surface perpendicular to direction of cut, non-randomness in orientation of grains exists. The grain and grain boundaries are elongated and oriented in the direction of cut, as shown in figure 5. The unaltered region further below is the bulk material zone which is unaffected by the machining conditions.

3.2. Quantitative analysis

Vickers microhardness measurements and grain orientation analysis are performed to quantify the depth and intensity of subsurface deformation.

3.2.1. Microhardness measurements. Vickers microhardness test was conducted on the study samples. The ploughing phenomenon of turning operation induces strain hardening in the near surface region which results in higher hardness values. From the figure 6, it is evident that the hardness values are higher near to the surface and gradually reduces reaching the bulk hardness.

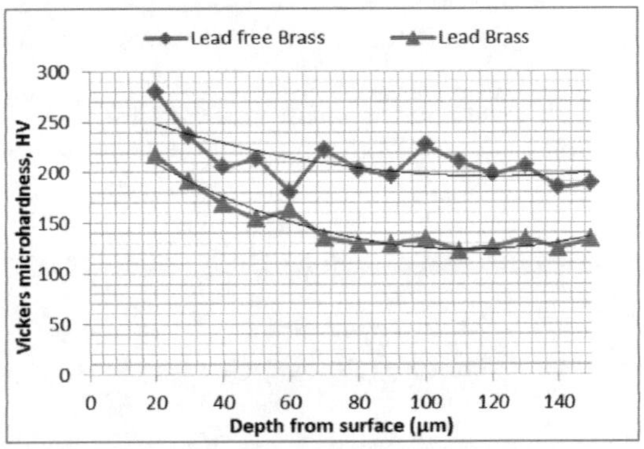

Figure 6. Vickers microhardness measurements

The hardness values of lead brass and lead free brass reach the bulk hardness values at approximately 80μm and 60μm respectively. Comparing the gradient of hardness values, higher hardness is induced by machining in lead brass compared to lead free brass.

3.2.2. Grain orientation. The average orientation of grains from image grid size 10μm x 250μm to 150μm x 250μm with an incremental size of 10μm in depth are measured and plotted as a function of depth from the surface. As earlier discussed, the grains are oriented randomly and even out to approximately 90 degrees in the bulk material zone. But the grains in the subsurface region of machined surface progressively orient towards the direction of cut and almost become parallel to the surface edge. The plots in figure 7 represent the grain orientation angle (θ) of lead- and lead free brass plotted as the function of depth from surface edge. The machining induced grain orientations are higher near the surface edge and lower in the region of bulk material.

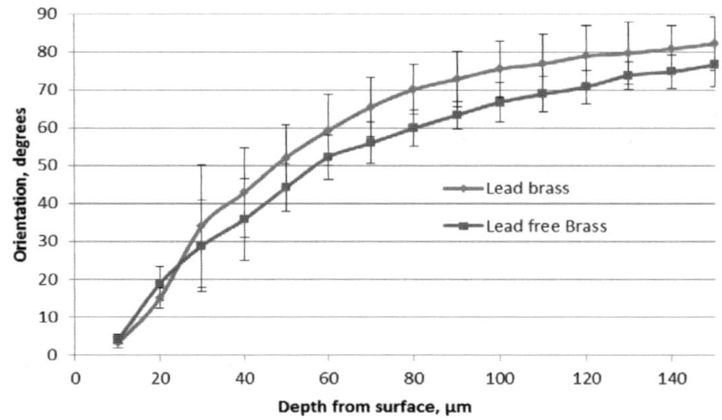

Figure 7. Variation of grain orientation of study samples as function of depth from surface edge

Further, the grain orientation angles are used to study the strain induced in the subsurface. Few investigations [18,19] on the flow lines in the subsurface region in a sliding test, estimated the equivalent strain, ε_{eq}, using the angle made by the flow lines as a function of depth from surface edge. CA Brown [20] calculated the strain induced in machined chips from the grain elongations. Similarly in this study, grain orientations in the subsurface region from ploughing effect of turning operation are used to calculate the equivalent strain, ε_{eq}, from (2).

$$\varepsilon_{eq} = \frac{\sqrt{3}}{3}\tan(90-\theta) \tag{2}$$

Where, θ is the grain orientation angle

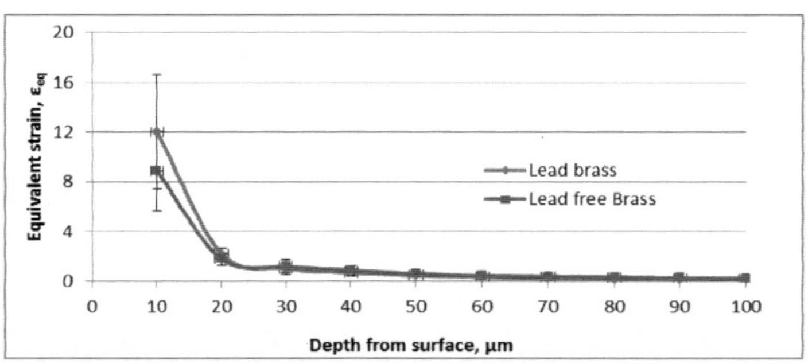

Figure 8. Equivalent strain as a function of depth from surface, calculated with mean grain orientations.

From figure 8, it is evident that the estimated plastic strain induced is higher in the subsurface region of study samples and the differences in the intensity of plastic strain in lead brass is marginal compared to lead free brass.

3.2.3. Underlying mechanism. Lead brass offers lower cutting forces at the tool/work interface, primarily due to the presence of lead, and as a result generates discontinuous chips, whereas lamellar chips are generated while machining lead free brass [1,13]. The formation of discontinuous chips while machining lead brass, due to lower tool forces, causes a ripping effect on the workpiece surface [21]. This ripping effect induces slightly higher roughness and additional strain than normal cutting effect on the lead brass surface.

IOP Conf. Series: Journal of Physics: Conf. Series **1183** (2019) 012006 doi:10.1088/1742-6596/1183/1/012006

4. Conclusion

The results and analysis are specific to the study samples and machining conditions. The following conclusions can be drawn from the study:

1. Vickers microhardness measurements suggest turning operation influence strain hardening in the subsurface region of the brass samples with higher induced hardness in lead brass.
2. The grain analysis suggest significant metallurgical alterations observed in the subsurface region perpendicular to the direction of cut and very little or insignificant grain deformation observed in the region perpendicular to the feed direction.
3. From the grain orientation analysis, effective strain is estimated and the differences in the plastic strain between the lead brass and lead free brass are found to be marginal.
4. Considering the hardness measurements and equivalent strain, intensity of subsurface deformation in turned lead free brass is in proximity with turned lead brass samples.

Therefore the study results suggest the possibility to attain subsurface characteristics of lead brass by replacing the lead content with silicon. Future work includes multi-scale investigations on subsurface region by comparing the results of grain orientations using electron microscopy with the crystal misorientation using Electron Back Scattered Diffraction (EBSD) technique.

Acknowledgment

The research presented in this paper is a part of the Lead Free Brass research project funded by MISTRA (the Swedish Foundation for Strategic Environmental Research), and SIO (the Swedish Strategic Innovation Area). The authors would like to thank; AB Markaryds Metallarmatur, Markaryd for their participation in this project; Digital Surf for providing us the software MountainsMap® Premium and all colleagues in the Functional Surfaces Research Group at Halmstad University for their help, support and cooperation.

References

[1] Trent E and P Wright 2000 *Metal Cutting*: *Fourth edition* Butterworth Heinemann Stoneham MA USA

[2] Vijeth V Reddy, Amogh Vedantha Krishna, Fredrik Schultheiss and B-G Rosén 2017 Surface Topography Characterization of Brass Alloys: Lead Brass (CuZn39Pb3) and Lead free Brass (CuZn21Si3P) Surf. Topogr.: Metrol. Prop. 5 025001 https://doi.org/10.1088/2051-672X/aa612b

[3] Guy Bellows, Dean N. Tishler 1970 Introduction to Surface Integrity General Electric Aircraft Engine Group Cincinnati Ohio

[4] V. I. Levitas, V. F. Nesterenko and M. A. Meyers 1998 Strain-Induced Structural Changes and Chemical Reactions Part II Acta materialia USA

[5] S.L. Rice, H. Nowotny, S.F. Wayne 1989 A Survey of the Development of Subsurface Zones in the Wear of Materials Key Engineering Materials Vol. 33 pp. 77-100

[6] D. Landolt, S. Mischler, M. Stemp, S. Barril 2004 Third Body Effects and Material Fluxes in Tribocorrosion Systems Involving a Sliding Contact Wear Volume 256 Issue 5 Pages 517-524 ISSN 0043-1648 http://dx.doi.org/10.1016/S0043-1648(03)00561-1

[7] A. Moshkovich, V. Perfilyev, I. Lapsker, D. Gorni, L. Rapoport 2011 The Effect of Grain Size on Stribeck Curve and Microstructure of Copper Under Friction in the Steady Friction State Tribology Letters 42 pp. 89-98

[8] A. Thakur, S. Gangopadhyay, K.P. Maity 2014 Effect of Cutting Speed and Tool Coating on Machined Surface Integrity of Ni-based Super Alloy Procedia CIRP Volume 14 Pages 541-545 ISSN 2212-8271 http://dx.doi.org/10.1016/j.procir.2014.03.045

[9] Mike Olsson, Volodymyr Bushlya, Jinming Zhou, Jan-Eric Ståhl 2016 Effect of Feed on Sub-surface Deformation and Yield Strength of Oxygen-free Pitch Copper in Machining Procedia CIRP Volume 45 Pages 103-106 ISSN 2212-8271 http://dx.doi.org/10.1016/j.procir.2016.02.163

[10] Z. Pan, Y. Feng, S.Y. Liang 2017 Material Microstructure Affected Machining: a Review Manufacturing Review 4 p. 5
[11] Fergani O, Liang SY 2015 The Effect of Machining Process Thermo-Mechanical Loading on Workpiece Average Grain Size Int J Adv Manuf Technol 80:21–29 doi:10.1007/s00170-015-6975-8
[12] Wieland-SW1 Lead-free Special Brass Wieland-Werke AG www.wieland.de 2014-01-28
[13] Wieland-Z32/Z33 Machining Brass Wieland-Werke AG www.wieland.de 2014-01-28
[14] Schultheiss F, Lundström E, Johansson D, Bushlya V, Zhou J, Nilsson K and Ståhl J-E 2014 Machinability of Lead-Free Brass—a Comparative Study Swedish Prod. Symp
[15] Canovic S, Jonsson T, Halvarsson M 2008 Grain Contrast Imaging in FIB and SEM Institute Phys. Conf. Series vol. 126 pp 012054
[16] ISO 1832:2012 2012–11 Indexable inserts for cutting tools—designation (International Organization for Standardization)
[17] ISO 6507-1:2005 Metallic materials -- Vickers hardness test -- Part 1: Test method
[18] Dautzenberg. J. H & Zaat J. H 1973 Quantitative Determination of Deformation by Sliding Wear Wear 23(1) 9-19 DOI: 10.1016/0043-1648(73)90036-7
[19] Pradeep L. Menezes, Kishore, Satish V. Kailas 2009 Role of Surface Texture of Harder Surface on Subsurface Deformation Wear Volume 266 Issue 1 Pages 103-109 ISSN 0043-1648 http://dx.doi.org/10.1016/j.wear.2008.05.008
[20] C.A. Brown 1987 Strain Analysis in Machining using Metallographic Methods Metallography Volume 20 Issue 4 Pages 465-483 ISSN 0026-0800 http://dx.doi.org/10.1016/0026-0800(87)90020-6
[21] Toenshoff H K and Denkena B 2013 Basics of Cutting and Abrasive Processes (Berlin: Springer)

IOP Conf. Series: Journal of Physics: Conf. Series **1183** (2019) 012007 doi:10.1088/1742-6596/1183/1/012007

Wear analysis of pin and roller surfaces

Z. El-Ghoul [1], S. John [1], Z. Dimkovski[1], P.-J. Lööf[1], J. Lundmark[2], J. Mohlin[3]
[1] Halmstad University, PO Box 823, SE-301 18 Halmstad, Sweden
[2] Applied Nano Surfaces AB, 12 Knivstagatan, 753 23 Uppsala, Sweden
[3] Gnutti Carlo Sweden AB, 60 Aröds Industriväg, 422 43 Gothenburg, Sweden

E-mail: zoeelg09@student.hh.se

Keywords: Tribology, friction, wear, surface topography, relocation

Abstract. In this paper three different combinations were evaluated: (i) steel pin vs steel roller, (ii) steel pin vs tungsten disulfide (WS_2) coated roller and (iii) PVD pin vs steel roller. The aim of this paper is to properly quantify the topographical variations of the pin/roller surfaces and asses the appropriateness of the measuring technique before and after the tribological tests primarily performed for friction reduction. In-house tribometer was used to test the friction/wear performance of the three different combinations. Surface topographies were measured by white light interferometer (3D) and mechanical stylus (2D). To enable fast relocation on the stylus and interferometer, a 3D printed mask was used. Regarding the pins, Sa, Sk and Svk have been found to follow the wear changes: the lowest wear change for the combination (ii) and highest for the combination (i) but less for the combination (iii). Regarding the changes of Ra on the pin locations where edge contact occurs, the largest changes were found to be for combination (i), followed by combination (ii) and (iii). Ra and Rk change less significantly for the roller surfaces. Good correlations among the friction, initial/final roughness and radial displacements were also found.

1. Introduction

Over the last decades, the government has imposed new regulations and standards requiring better fuel economy and lower emissions for heavy-duty diesel engines. At low engine speeds, valve train friction makes a significant contribution to the engine friction [1], especially in the pin-roller assembly, which adversely affects the durability of these components [2]. These regulations became a necessity to drive the product development. Some of these demands are directly related to friction, wear and lubrication of these mechanical components depending on the operating conditions and the original surface topography [3, 4]. Wear plays an important role in determining life span of products of machine elements. By using wear resistance materials, the wear becomes smaller and down to nanometer scale. Roughness has a strong impact on wear in valvetrain components [5]. The usage of coatings is growing strongly; a huge progress has been achieved in the development of surface treatments. It has been found that the tungsten disulphide (WS_2) coating materials have reduced the friction wear operating at low speeds and high loads. The main reason for using the WS_2 coating is the cost efficiency of the process called "triboconditioning" introduced by Applied Nano Surfaces Sweden AB. It combines burnishing with deposition of a low-friction film of tungsten disulphide on the surface compared to the uncoated materials WS_2 shows higher temperature [6-10]. The PVD has been used very successfully to improve fatigue resistance [11] and improving the lifetime with larger friction reduction of rolling contacts under high contact pressure [12, 13]. But it has found that at low speed

IOP Conf. Series: Journal of Physics: Conf. Series **1183** (2019) 012007 doi:10.1088/1742-6596/1183/1/012007

and low pressure of the sliding contact, the WS_2 soft coating also improves the performance instead of more expensive PVD hard coating. This is probably because: (i) a better surface finish with flatter asperities, (ii) a low-friction tribofilm and (iii) a low wear-reduction, especially in lubricated sliding contacts at the boundary and mixed regime [6-10]. There are other di-sulphide coatings like MoS_2 also governing the tribological performance, but at low sliding speed the WS_2 coating improves the performance because of its thermal stability and less sensitivity to humidity compared to the MoS_2 coating [10]. Nowadays, there are still attempts to improve tribological properties through the methods of surface engineering and better understanding of wear mechanisms in relation to the evaluation of the surface condition [14]. Surface topography measurement can be used for characterization of types of wear and for microscopic wear measurement. There are many ways to measure and evaluate wear. In some cases, wear amount can be obtained after the measurement of worn surface by using relocation techniques [15, 16]. A commonly used technique is to measure wear tracks by using 2D stylus profilometer [17] or 3D interferometer [18].

The aim of this paper is to properly analyze the topographical variations of the pin/roller surfaces before and after the tribological tests primarily performed for evaluation of friction reduction. Another goal is to analyze the wear scratches with a proper relocation technique [19]

2. Materials and Methods

2.1. Surface samples

The coated samples are considered to have the same macro geometry before and after surface treatment. The initial profile analysis is performed and presented in table 1. Each combination consists of five samples of pins/rollers. All pins have the same dimension with 25 mm in diameter and 55 mm in length. Roller has standard dimension of 35 mm in diameter and 25 mm in length.

Table 1. Sample combinations

Combinations	(i) steel pin vs steel roller	(ii) steel pin vs tungsten disulfide (WS_2) coated roller	(iii) PVD coated pin vs steel roller
Pins			
Initial profile	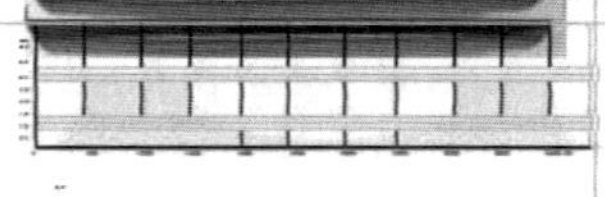	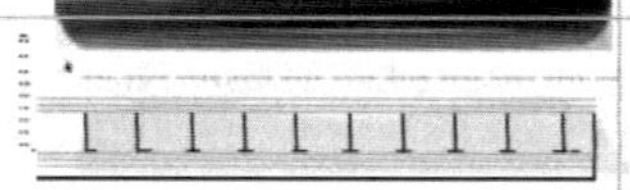	
Ra and Rvk	Ra=0.031μm. Rvk=0.052μm	Ra=0.028μm. Rvk=0.047μm	Ra=0.037μm. Rvk=0.08μm
Rollers	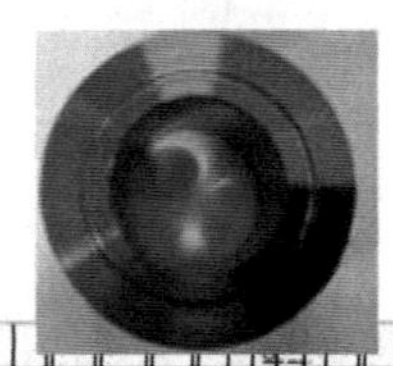		
Initial profile	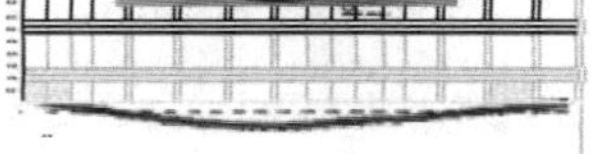	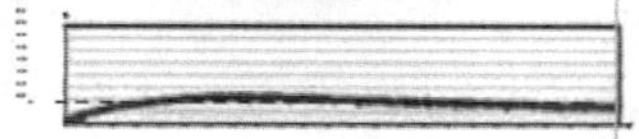	
Average of Ra and Rvk	Ra=0.058μm. Rvk=0.085μm	Ra=0.031μm. Rvk=0.06μm	Ra=0.06μm. Rvk=0.088μm

2.2. In house tribometer

A dedicated test rig for pin-roller contact testing has been developed in-house (see figure1.a) in order

IOP Conf. Series: Journal of Physics: Conf. Series **1183** (2019) 012007 doi:10.1088/1742-6596/1183/1/012007

to test the standard size components under controlled conditions. The experimental setup and schematic of tribometer illustrated in figure 2.b. Instead of an eccentric cam as in a real engine, a cylindrical cam driven by the electric motor-A is used to drive the roller under a constant load in order to stabilize the load, increase the accuracy and repeatability, a suspension with flexible steel sheets between the pin-roller assembly and the load sensor was used. The load is adjusted by moving the whole table vertically by the electric motor-B. Electrical heating elements connected to a control system maintain the oil at a stable temperature and an oil recirculation system keeps the oil pressure at a correct level. Data-acquisition and speed control of motor-A is achieved by LabView [20].

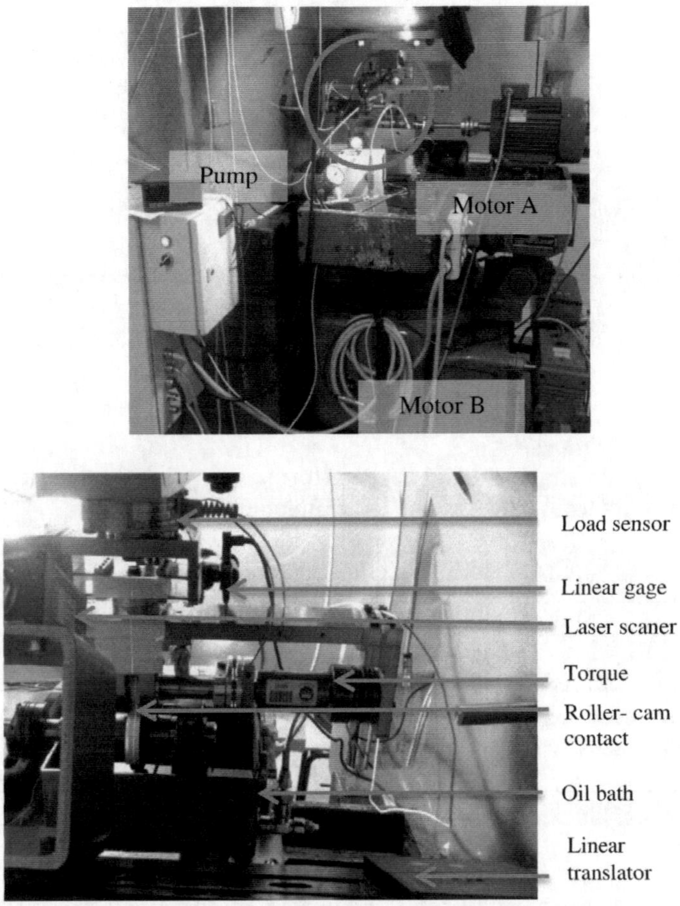

Figure 1a. Tribometer set up used at Halmstad University.

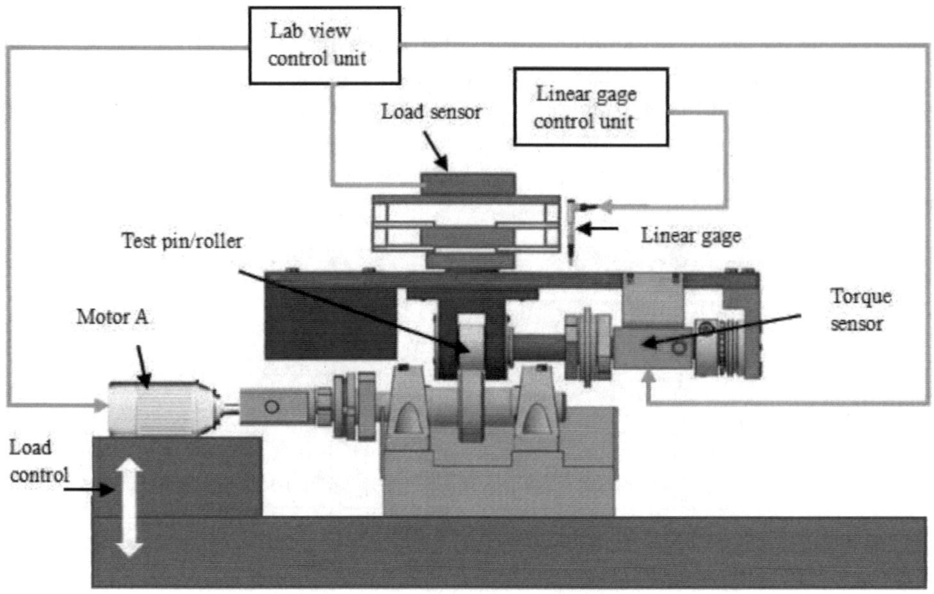

Figure 1b. Schematic of in-house tribometer setup

2.3. Running conditions and procedure

Before performing any test, an alignment had been performed and discussed in [21]. A summary of the initial conditions /parameters for the tests is presented in table 2.

Table 2. Rig test parameter

Test Conditions	Value
Load	**5 kN**
Oil. Temperature in path/inlet to contact	**38°C**
Oil viscosity /pressure	**SAE 0W-20 at 1 bar**
Rotational speed	**20-1130 rpm**

2.4. Measuring methodology and Surface characterization

Before conducting the roughness measurements, a contact configuration performed to predict the zone area of contacts between pin/roller and the driving cam, the configuration will help to relocate the area of the contact between the mating surfaces. A measuring methodology used for evaluating changes between unworn and relocated worn surface, 3d printed mask was used to enable fast relocation for the pins and rollers. This measuring method can be used typically for all the testing samples. Due to the limitations only the 2D measurements were conducted in the inner surface of the roller.

The 2D measurements were taken at three angular positions: 72 °, 90° and 108° each cylinder from of each pin and four angular sectors (0º. 90º. 180º .270º) on the roller surfaces as shown by the figure 2b. Measurements taken with the help of Somicronic Surf scan 3CS mechanical stylus system (resolution of 1 µm m in x. 6 nm in z-direction). A probe with 2 um tip radius and 90̊ tip angle were used for measuring pin/roller surfaces running with a speed of 0.3 mm/s.

IOP Conf. Series: Journal of Physics: Conf. Series **1183** (2019) 012007 doi:10.1088/1742-6596/1183/1/012007

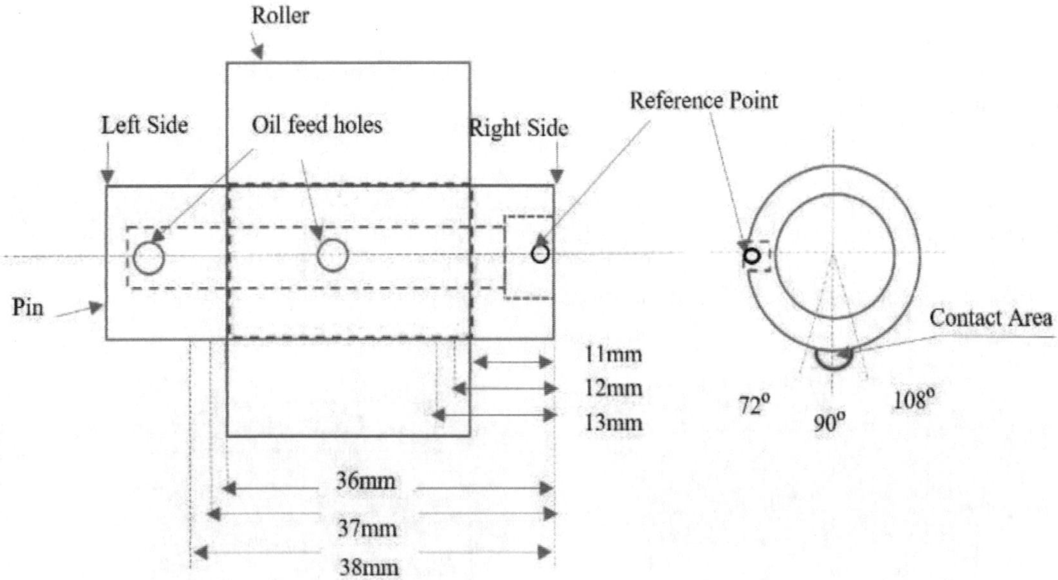

Figure 2a. Roller and pin contact area and profile measurement for the pin.

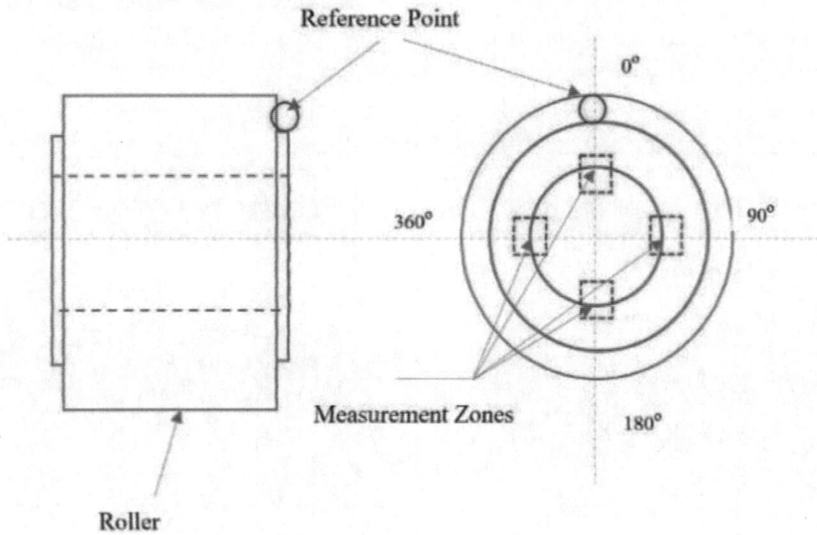

Figure 2b. Profile measurement positions over the roller

The 3D measurements were taking directly on the surfaces 6 positions (11. 12. 13. 36. 37 38mm) from the left side of each pin direction) in the axial direction of the pin are chosen as shown by the figure 2a. Six positions/regions were measured with 10 X magnifications (600 µm x 800 µm measurement area. 0.1nm vertical resolution) with non-contact white light interferometer Phase shift MicroXam. The data were analyzed in the Mountainsmap v7 software, removing the second order polynomial. and then filtered with 3D high pass Gaussian filtering with a cut-off length of 0.25mm, the 3D standard parameters are included this study from are height parameter and functional

IOP Conf. Series: Journal of Physics: Conf. Series **1183** (2019) 012007 doi:10.1088/1742-6596/1183/1/012007

parameters according to ISO 25178-2.

3. Results

The results will be presented in two parts, surface measurements techniques and correlation study the result of the friction behaviour presented in [21]

3.1. Relocation

Figure 3 shows the surface topographies of the three different combinations before and after the test. The roughness calculated from the extracted profile and results show that steel pins run against the tungsten disulfide (WS2) coated roller (Sa=0.48µm. Sk=0.099µm. Svk=0.98µm) have lower wear than pins run versus the steel roller (Sa=0.53µm. Sk=0.161µm. Svk=0.101µm.

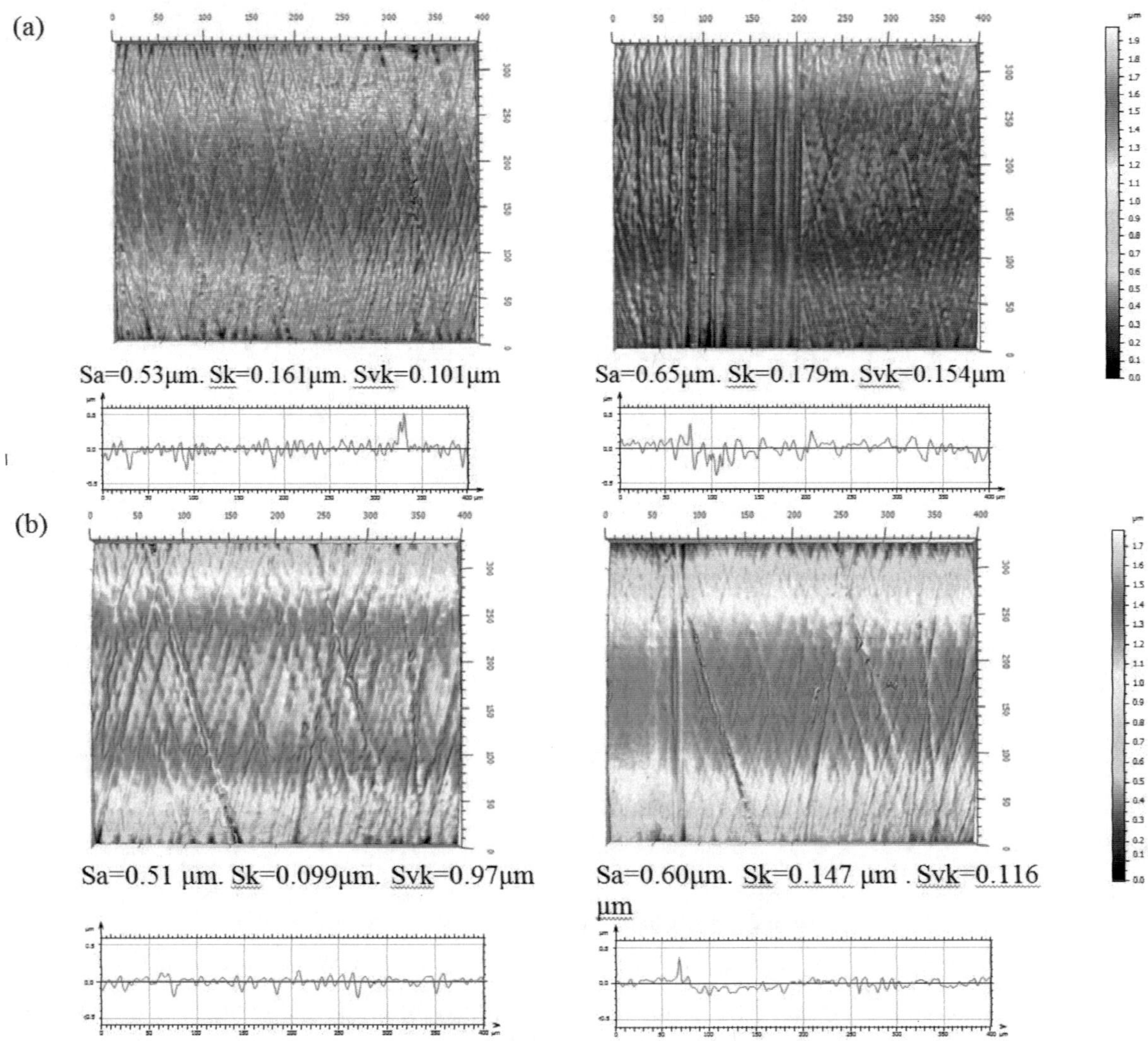

IOP Conf. Series: Journal of Physics: Conf. Series **1183** (2019) 012007 doi:10.1088/1742-6596/1183/1/012007

(c)

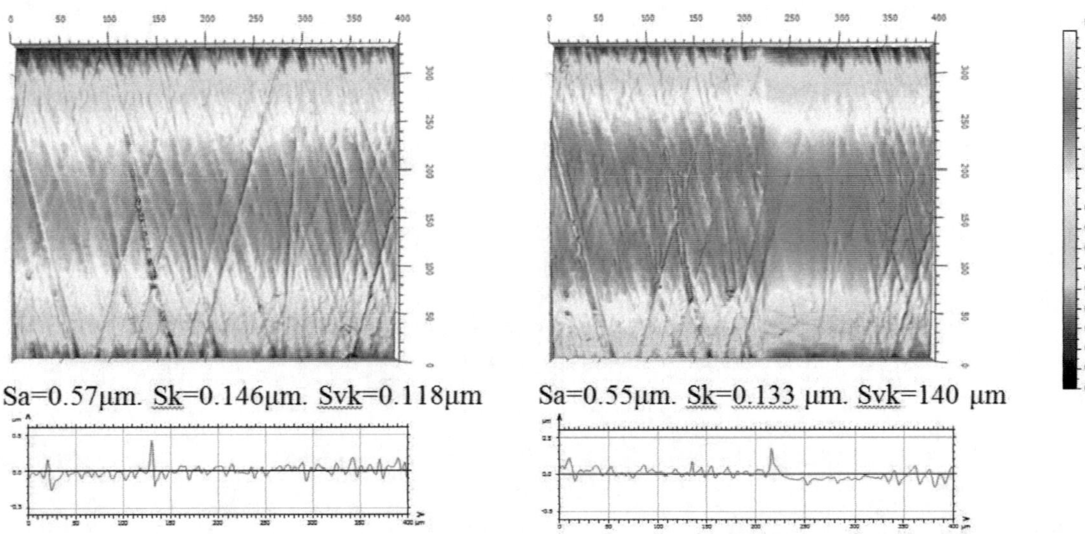

Figure 3 3D surface topographies 2D profile of wear tracks of the three combinations before and after the test (a) combination (i) steel pin vs steel roller showing higher peak and valley before the test, (b) combination (ii) steel pin vs tungsten disulfide showing less peak and valley before the test,(c) combination (iii) PVD pin vs steel roller show more roughness and more valley before the test.

3.2. 2D Roughness parameters

Figure 4 (left) shows the Ra of the relocated surfaces evaluated separately on the left and right side of the pin (averages of 40 measurements). For the steel pin combination, Ra on the left side is higher than of the left side after the test while the WS_2 combinations show lower values of right side rather than the left side.

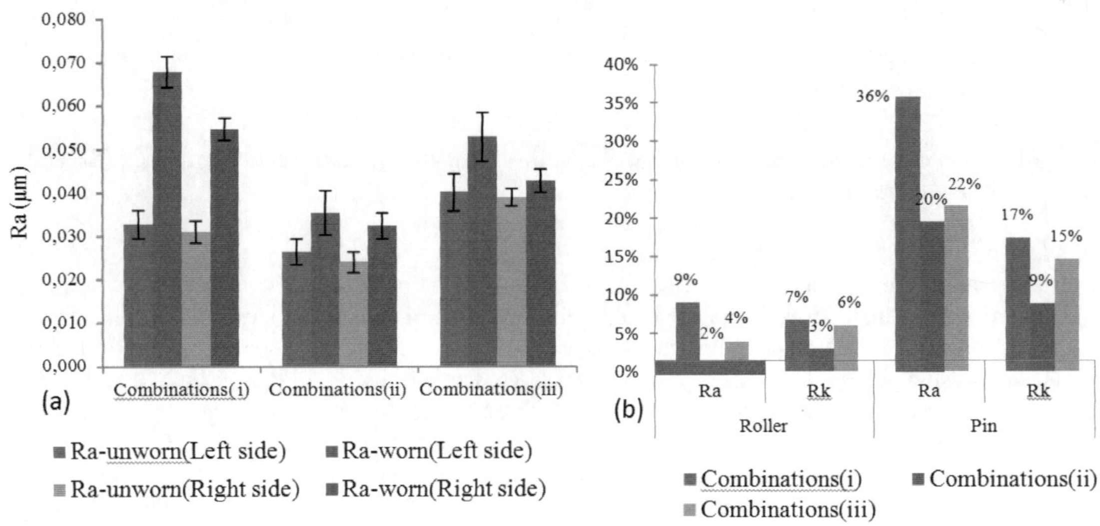

Figure 4. (a) Ra of pins evaluated at the left and right locations. (b) Percentage of Ra and Rk changes for the rollers and pins. The parameters are averages evaluated at the left/right locations on the pins and at four angular positions on the rollers.

Figure 4 right shows the Ra and Rk changes for the rollers and pins because Ra and Rk were the most stable parameters. 60 measurements were evaluated for the roller. Ra and Rk change less significantly for the roller surfaces. The Ra and Rk for the pin surfaces show that the steel pin vs WS_2 has the lowest changes while the steel pin vs steel roller has the highest one. The PVD pins do not change much on the right side after the test.

3.3. 3D Roughness parameters analysis

Figure 5 left show the variation of Sa, Sk and Svk which are the averages of five 10x interference measurements of the three different combinations. The total percentage of increase after the test is calculated by: Percentage of increase (%) = (Sx before – Sx after)/ Sx before * 100. where Sx = Sa, Sk or Svk. It can be noticed that the parameters seem more stable and clearly show that WS_2 type changes less (Sa=7%, Sk=10% and Svk=16%) and steel type changes more (Sa=25%, Sk=12% and Svk=41%).

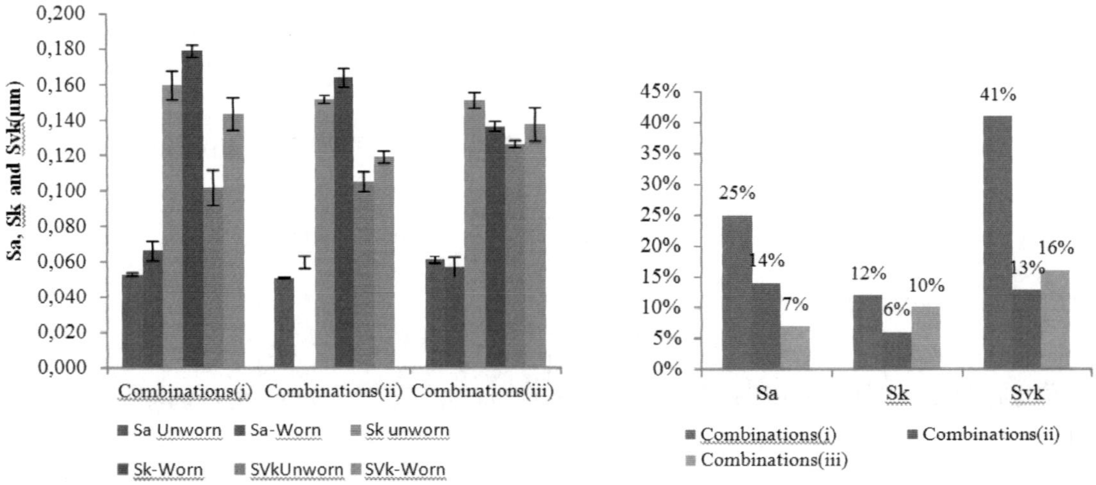

Figure 5. Left: the variations of Sa, Sk and Svk for the three different pin types. Right: Percentage of Sa, Sk and Svk changes for the pins.

The figure 5 right shows the variation of Sa, Sk and Svk for the three different pin types (Sk and Svk show more clear differences).

3.4. Correlation

The linear correlation coefficients among the rollers 2D roughness parameters, coefficient of friction-COF and radial displacement (total radial displacement Y and local radial displacement amplitude y) were calculated (see table 3). Values only over 0.9 are considered significant. Strong correlation exists among the radial displacement, COF and the parameters: Ra, Rq, Rp, Rz, Rdc, Rku, Rdc, Rk and Rvk. Figure 6 shows the radial displacement of the three different combinations which were tested under same conditions which is the average of 50 readings of each combinations.

The steel type combinations (i) had the largest total radial displacement (Y) and the largest local radial displacement amplitude (y) followed by the PVD and WS_2 type. Vertical displacement amplitudes were measured by using linear gage Mitutoyo LGB2 and recorded with SENSOPARK software.

IOP Conf. Series: Journal of Physics: Conf. Series **1183** (2019) 012007 doi:10.1088/1742-6596/1183/1/012007

Figure 6. The displacement amplitude of three combinations with global(Y) and local(y) measurement.

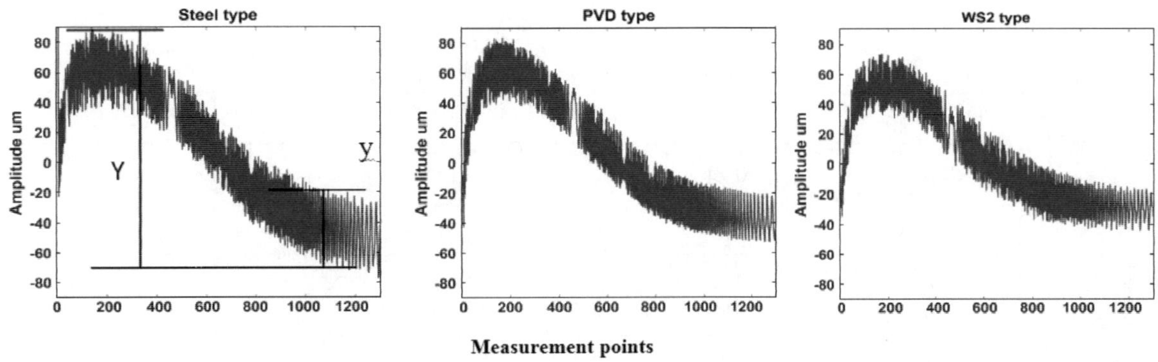

Table 3 the correlation coefficient matrix among the rollers 2D roughness parameters, radial displacement, and coefficient of friction-COF at three different speeds (20 rpm, 40 rpm and 60 rpm). The 2D roughness parameters of pins are the average of 15 readings and for roller 20 readings. COF friction calculated from the average of 50 readings and radial displacement mentioned in average of 50 readings.

Values only over 0.9 are significant. Only the significant values are shown in the tables. The positive values mean the parameter and the response have the same variation.

IOP Conf. Series: Journal of Physics: Conf. Series **1183** (2019) 012007 doi:10.1088/1742-6596/1183/1/012007

Table 3. The correlation coefficients matrix among the rollers 2D roughness parameters, radial displacement, and coefficient of friction-COF at three different speeds (20 rpm. 40 rpm and 60 rpm).

		COF 20rpm	COF 40rpm	COF 60rpm	Global vertical displacement	Local vertical displacement
Coefficient of friction	COF 20rpm					
	COF 40rpm	0.99				
	COF 60rpm	1	0.98			
2D parameter Unworn	Global vertical displacement	0.99	1	0.97		
	local vertical displacement	1	0.99	1	0.98	
		0.88	0.93	0.85	0.95	0.86
	Rq	0.88	0.93	0.85	0.95	0.86
	Rp	0.89	0.94	0.86	0.95	0.87
	Rv	0.85	0.90	0.81	0.93	0.83
	Rz	0.88	0.92	0.84	0.94	0.85
	Rdc	0.88	0.93	0.85	0.95	0.86
	Rk	0.88	0.93	0.85	0.95	0.86
	Rpk	0.89	0.94	0.86	0.96	0.87
	Rvk	0.86	0.91	0.82	0.93	0.83
2D parameter worn	Ra	0.99	1	0.99	1	0.99
	Rq	1	1	0.99	0.99	0.99
	Rp	0.96	0.92	0.98	0.9	0.97
	Rv	0.98	0.99	0.96	1	0.97
	Rz	1	0.98	1	0.97	1
	Rdc	0.99	1	0.98	1	0.98
	Rk	0.99	1	0.98	1	0.98
	Rpk	0.93	0.89	0.96	0.86	0.95
	Rvk	0.99	1	0.98	1	0.99

Table 4 shows the correlation coefficients matrix among the pins 2D/3D roughness parameters, radial displacement, and coefficient of friction-COF at three different speeds. The 2D roughness parameters of pins are the average of 15 readings and 20 readings for roller. The 3D parameters which are the average of 10 readings show significant correlation after the test. Rk has strong correlation with COF for the unworn surfaces while Sk and Svk correlate well with COF and radial displacement before and after test. Values only over 0.9 are significant. Only the significant values are shown in the

IOP Conf. Series: Journal of Physics: Conf. Series **1183** (2019) 012007 doi:10.1088/1742-6596/1183/1/012007

tables. The positive values mean the parameter and the response have the same variation. While the negative sign means the parameter and the response have the different variation

Table 4. The correlation coefficients matrix among the pins 2D/3D roughness parameters, radial displacement and coefficient of friction-COF at three different speeds (20 rpm 40 rpm and 60 rpm).

Pin		COF 20rpm	COF 40rpm	COF 60rpm	Global vertical displacement	Local vertical displacement
COF	COF 20rpm					
	COF 40rpm	**0.99**				
	COF 60rpm	**1**	**0.98**			
	Global vertical displacement	**0.99**	**1**	**0.97**		
	Local radial displacement Amplitude	**1**	**0.99**	**1**	**0.98**	
Pin 2D parameter Unworn	Ra	0.47	0.56	0.40	0.60	0.43
	Rz	0.32	0.42	0.25	0.47	0.27
	Rc	0.39	0.49	0.32	0.53	0.35
	Rku	0.12	0.22	0.05	0.28	0.07
	Rdc	0.47	0.56	0.41	0.61	0.43
	RPc	-0.12	-0.22	-0.05	-0.28	-0.07
	Rk	**0.99**	**0.96**	**1.00**	**0.94**	**0.99**
	Rvk	0.12	0.22	0.05	0.28	0.07
Pin 2D parameter Worn	Ra	**0.97**	**0.94**	**0.98**	**0.91**	**0.98**
	Rz	**0.96**	**0.93**	**0.98**	**0.91**	**0.97**
	Rc	**0.98**	**0.96**	**0.99**	**0.94**	**0.99**
	Rku	**0.97**	**0.94**	**0.99**	**0.92**	**0.98**
	Rdq	**0.96**	**0.92**	**0.98**	**0.9**	**0.97**
	Rdc	**0.99**	**1**	**0.98**	**1**	**0.99**
	RPc	**-0.94**	**-0.9**	**-0.96**	**-0.87**	**-0.95**
	Rk	**0.96**	**0.99**	**0.94**	**0.99**	**0.95**
	Rvk	0.81	0.87	0.85	0.87	0.84
Pin 3D Parameter unworn	Sa	0.62	0.70	0.57	0.74	0.59
	Sq	0.34	0.44	0.27	0.49	0.29
	Sz	0.74	0.66	0.78	0.62	0.77
	Sk	**0.97**	**0.99**	**0.95**	**1.00**	**0.95**
	Svk	**0.92**	0.87	**0.94**	0.84	**0.93**
Pin 3D Parameter worn	Sa	**1.00**	**0.98**	**1.00**	**0.97**	**1.00**
	Sq	**1.00**	**0.99**	**1.00**	**0.98**	**1.00**
	Sz	**1.00**	**0.98**	**1.00**	**0.97**	**1.00**
	Sk	**1.00**	**1.00**	**0.99**	**0.99**	**0.99**
	Svk	**0.95**	**0.98**	**0.93**	**0.99**	**0.94**

4. Discussion

The roughness analysis of pin and roller were measured by stylus and interferometer. 2D measurements were significant while measuring inner surfaces of the roller. 3D measurements were significant to trace the wear scratches over the pin surface. To enable fast relocation on the stylus and interferometer, a 3D printed mask was used. The roughness parameters were selected for this study according to the significant changes in the correlation matrix see table 3 and 4. WS2 coated roller surfaces showing lower roughness than than both rhe PVD type and steel type (see figure 4 and 5). 60 measurements were evaluated for the roller. 45 measurements were evaluated for the pin. Ra, Rk and Rvk in 2D and Sa, Sk and Svk in 3D change more significantly for the roller and pin surface.

The results of combination (ii) samples reduce the rolling wear; it can be seen in (figure 3.b). The Tungsten disulphide (WS_2) roller coatings and the burnishing effect enabled smoother surface and enhances the wear resistance. On other hand vertical displacement behavior of each combination follow the trend of friction and roughness. Indeed, the relation of initial roughness with friction behaviour and vertical displacement behaviour illustrated in table 3 and table 4. The roughness of pin/roller surfaces are highly correlated with the coefficient of friction.

5. Conclusion

- The roughness increase after test and the pin surface shows more significant change than the roller suface. The Ra of the pin surfaces increase in the following order WS_2 (20%), PVD (22%) and steel type (35%). Ra of the roller surfaces increase as: WS2 by 2%, PVD by 4% and steel type by 7%.
- The roughness value Ra of pin over left side of measurements is higher than right side. The trend has the following decreasing order: WS_2 coated type < PVD type < steel type. Another fact is that the initial roughness of PVD pin () is higher than WS_2 pins (Ra = 0.026 μm).
- Ra, Rk, Rvk, Sa, Sk and Svk parameters are more stable for evaluation.
- Strong linear correlation exists between the radial displacement, COF and the 2D parameters Ra, Rq,Rp Rz, Rdc, Rku, Rdc, Rk and Rvk for the roller surfaces . While for the pin surfaces Rk has strong correlation with COF for the unworn surfaces. Sk and Svk show significant correlation with COF and radial displacement before and after test.
- In [21] was found out that the friction was highest for the steel combination, medium for the PVD combination and lowest for the WS_2 combination. The wear manifested in *Sa* and *Svk* changes of the pins follow this trend of the coefficient of friction.

Acknowledgements

The authors would like to thank to the Vinnova and DigitalSurf for providing the MountainsMap software.

References

[1] Calabretta M. Cacciatore. D and Carden P (2010) *Valvetrain Friction - Modeling. Analysis and Measurement of a High-Performance Engine Valvetrain System*. SAE Int. J. Engines 3 2 pp 72-84. doi:10.4271/2010-01-1492

[2] Y. C.Tasan Y.C, *J2005 Measurement of Deformation in Rolling and Sliding Contacts (Ph.D.Thesis) University of Twentey. Enschede. the* Netherlands.

[3] Nilsson R, 2005, *On wear in rolling/sliding contacts (Ph.D. thesis*). Royal Institute of Technology KTH

[4] Zhang J, S2006 Detection and monitoring *of wear using imaging methods*. University of Twente. Enschede. the Netherlands.

[5] Cabanettes . F. Claret-Tournier. J. Mohlin. J.Nilsson. P.H. Rosén. Xiao. L, 2007 "The evolution of the surface topography of injection cams". 11th International Confernece of Meterology and Properties of Engineering surfaces. Huddersfiled. UK(2007)

[6] Zhmud B 2011 Tribology and Lubrication Technology 67 42

[7] Zhmud B 2012 Proc. STLE Annual Meeting (St. Louis. USA. May 7-10

[8] Zhmud B. Kolar C. Patschull J. Morawitz U and Broda M 2012 Proc. VDI-Conf. Ventiltrieb und Zylinderkopf (Wurzburg. Germany. November 27-28)

IOP Conf. Series: Journal of Physics: Conf. Series **1183** (2019) 012007 doi:10.1088/1742-6596/1183/1/012007

[9] Zhmud B. Tomanik E and Xavier F-A 2014 Lubrication Science 26 277
[10] Dimkovski Z.. Guilbert F Lundmark J Mohlin J Rosén B.-G.. (2015). *Optimization of theTriboconditioning Process on External Cylindrical Surfaces*. in Proceedings of the 15thInternational Conference of Metrology and Properties of Engineering Surfaces. March 2-5. 2015. Charlotte. USA.
[11] Rosado L, Jain V and Trivedi H, 1997, *The effect of diamond-like carbon coatings on the rolling fatigue and wear of M50 steel*, Wear, volume 212: page1–6.
[12] Kano M, 2006, *Super low friction of DLC applied to engine cam follower lubricated with ester-containing oil,* Tribology International 39 (2006) 1682–1685.
[13] Klaffke D, Santner E and Spaltmann D, 2005, *Influences of the tribological behaviour of slip-rolling DLC-coatings*, Wear, Volume 259, Issues 1–6, Pages 1-800.
[14] Chassaing G., Faure L., Philippon S., Coulibaly M., Tidu A., Chevrier P., Meriaux J., Adhesive wear of a Ti6Al4V tribopair for a fast friction contact, Wear, Vol. 320. 2014, p. 25–33.
[15] Wang W and Wong PL,V *Wear volume determination during running-in for PEHL contacts. In Tribology International*. Volume 33. Issue 7.. Pages 501-506. ISSN 0301-679X. https://doi.org/10.1016/S0301-679X(00)00091-8.
[16] F. Cabanettes, 2011, *Cam Roller Contact: Surfaces and Tribology (Ph.D.thesis),* Chalmers University of Technology, Gothenburg, Sweden.
[17] S. Johansson. P.H.Nilsson. R.Ohlsson. C.Anderberg. B.-G.Rosén. *New cylinder liner surfaces for low oil consumption.* Tribol.Int.41 (2008)854–859.
[18] D.J. Whitehouse, 1980. *The effect of surface topography on wear, fundamentals of tribology, SuhandSaka,* MITPress, Cambridge, USA, 1980.
[19] F. Cabanettes, B.-G. Rosén 2014, *Topography changes observation during running-in of rolling contacts,* Wear, vol. 315, pp, 78-86
[20] LABVIEW LabVIEW (64-bit), http://www.nationalinstrument.com version 8.3.0.532 (R2014a). 64-bit. February 2014.
[21] John S. El-Ghoul Z. Dimkovski Z. Lööf P-J. Lundmark J and Mohlin M (2017) *Friction between Pin Roller of a Truck's Valvetrain in:* Proceedings of the 16th International Conference on Metrology and Properties of Engineering Surfaces. 27-29 June 2017 Gothenburg. Sweden
[22] MountainsMap Premium. v7, http://www.mountainsmap.com.1.7037. 2014/04/08.

IOP Conf. Series: Journal of Physics: Conf. Series **1183** (2019) 012008 doi:10.1088/1742-6596/1183/1/012008

Identification of the minimum chip thickness based on the two-dimensional wavelet analysis

W Makieła, D Gogolewski and Ł Nowakowski
Kielce University of Technology, Faculty of Mechatronics and Mechanical Engineering, Department of Manufacturing Engineering and Metrology, 25-314 Kielce, Poland

E-mail: dgogolewski@tu.kielce.pl

Abstract. In this paper, authors take attempt to adapt the two-dimensional wavelet transform to assess the value of the minimum chip thickness for the face milling process of C45 steel. The images of surface layer from selected measuring devices were analyzed using a mother wavelet *bior6.8*. The analysis based on the wavelet transform revealed in signals three characteristics zones of the impact of cutting insert on the workpiece during initiation of the cutting process. The obtained values allowed to determine the relative method accuracy and to evaluate the possibility of application the particular measuring machines to assess the value of h_{min} parameter.

1. Introduction
The quality of the surface layer of the manufactured parts has significant influence on their mechanical properties [1,2]. The finishing process is the final step of the production process, which improve the surface texture of the manufactured parts. One of the most important parameters of milling process, during treatment with small depth of cut is the evaluation of the minimum chip thickness, which is the border value defining the beginning of machining process and the removal of material in chip form. Depending on the depth of cut it can be distinguished three zones of the impact of cutting insert on the workpiece. The elastic deformation and plastic kneading/crushing of workpiece occurs in the first zone. In this case, the value of depth of cut is lowest than the value of material which can be removed in the process ($a_p < h_{min}$). In the next zone, elastic-plastic deformation of the partial cutting process of workpiece takes place ($a_p \approx h_{min}$). Material removal in form of the chip occurs in the last zone ($a_p > h_{min}$) [3].

Analyzing the current state of art it can be noted that it is possible to assess the h_{min} value using analytical formulas [4-6] as well as using practical solutions, which allow to experimentally assess parameter value [7]. In analytical models are used relationship between cutting parameters, tool and workpiece. However, these models assume some simplification and idealization of process, which in real terms is not feasible. Empirical solutions are alternative to analytical approach. The h_{min} value can be evaluate based on analysis of the characteristic contact zones during initiation of the cutting process using microhardness analysis or optical and contact profilometer.

2. Two-dimensional wavelet transform
Many optical devices can be employ to measure the surface topography of machine parts. However, not all of them provide proper resolution in the particular axes, so that it is not possible to determine the locus of the particular contact zones during initiation of the cutting process. There are many methods

IOP Conf. Series: Journal of Physics: Conf. Series **1183** (2019) 012008 doi:10.1088/1742-6596/1183/1/012008

which can be apply to analyze signal in order to detect in them the characteristic information. One of this method is wavelet analysis which was developed at the turn of the twentieth and twenty-first century. Wavelet transform is mainly used to analysis of non-stationary signals, that is, those which are characterized by non-periodic irregularities. The application of wavelet transform lead to removing background features. It allow to detect characteristic information such as drifts, trends or abrupt signal changes and determine the location of such information in the signal [8].

The wavelet transform can be apply to 2D and 3D signal processing in the field of mechanical engineering. Many researchers have used wavelet transform to analyze the GPS signals [9-16]. In the work [9] were used a engineering and biomedical surfaces to demonstrate the application of wavelet analysis in the assessment of surface topographical features. The application 2D continuous wavelet transform to study the effect of abrasive finishing were presented in [10]. In [11] authors applied wavelet transform to filtering of freeform surfaces. Multiscalar wavelets were used to analyze the surface topography of wear of orthopaedic implants in [12]. In the work [13] were analyzed the deviations of cylindrical surfaces. In order to assess the tool flank wear the turned surface images were analyzed in [14]. In the work [15] authors applied wavelet analysis for detecting low-contrast defects under various light conditions in magnetic tile images. The work [16] present results of the application of wavelet transform to assess nanotopography of crystal surfaces.

Discrete two dimensional wavelet decomposition is based on the application of two filters, high pass and low pass. Therefore, at each decomposition level, the input signal is divided into four different signals: horizontal, vertical, diagonal details and approximation. Thus, it is possible, on particular level of analysis to detect characteristic information, which is not visible at other decomposition level [17].

3. Methodology and research results

In order to evaluate the possibility of application the two-dimensional wavelet analysis to assess the value of the minimum chip thickness were prepared the cuboid samples. The key issue was to produce the samples which surface was as smooth as possible, therefore all samples were polished and then the flatness of surface was measured. The research was carried out on the Prismo Navigator measuring machine using the VAST GOLD S-ACC probe head. In table 1 are presented the research results of the surface flatness. Thereafter the samples were mounted on inclined prism and there were milled. The process was based on the flowing change of depth of cut. The vertical milling centre AVIA VMC 800 was employed to carry out cutting process. The scheme of milling process is shown in figure 1. The process was performed with specified milling parameters v_c = 250 m/min and f_z = 0,1 mm/tooth. In order to avoid axial and radial run-out, one cutting insert was used. In the research was used cutting insert SDMT 120508 SN-81. The cutting process was performed without the cooling and lubricating liquids.

Table 1. The research results of the surface flatness

No.	surface flatness (µm)
1	1.3
2	1.3
3	2
4	2
5	1.3

IOP Conf. Series: Journal of Physics: Conf. Series **1183** (2019) 012008 doi:10.1088/1742-6596/1183/1/012008

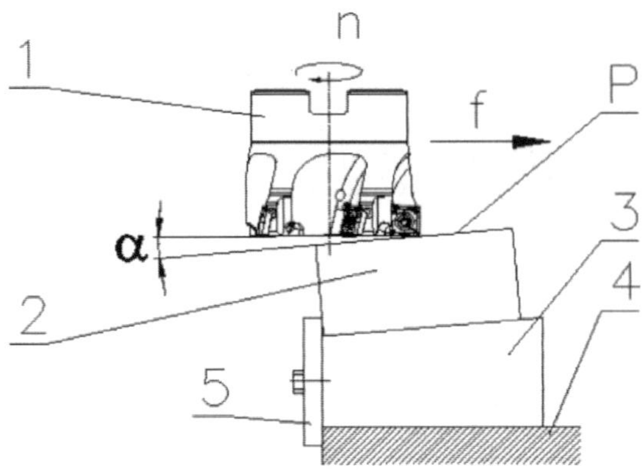

Figure 1. View of milling process. 1- tool, 2- sample, 3- prism, 4- handle, 5- stopper, P- working surface of the sample, f- feed, n- rotation, α- angle of working surface

Measurements were performed using the following measuring machines: optical profilometer Talysurf CCI, multisensor measuring machine O-Inspect 442 and workshop microscope MarVision MM320 using three levels of magnification (x0,7, x1, x2). The optical profilometer was adopted as a reference device. The selected images of surface for particular measuring devices are presented in figure 2.

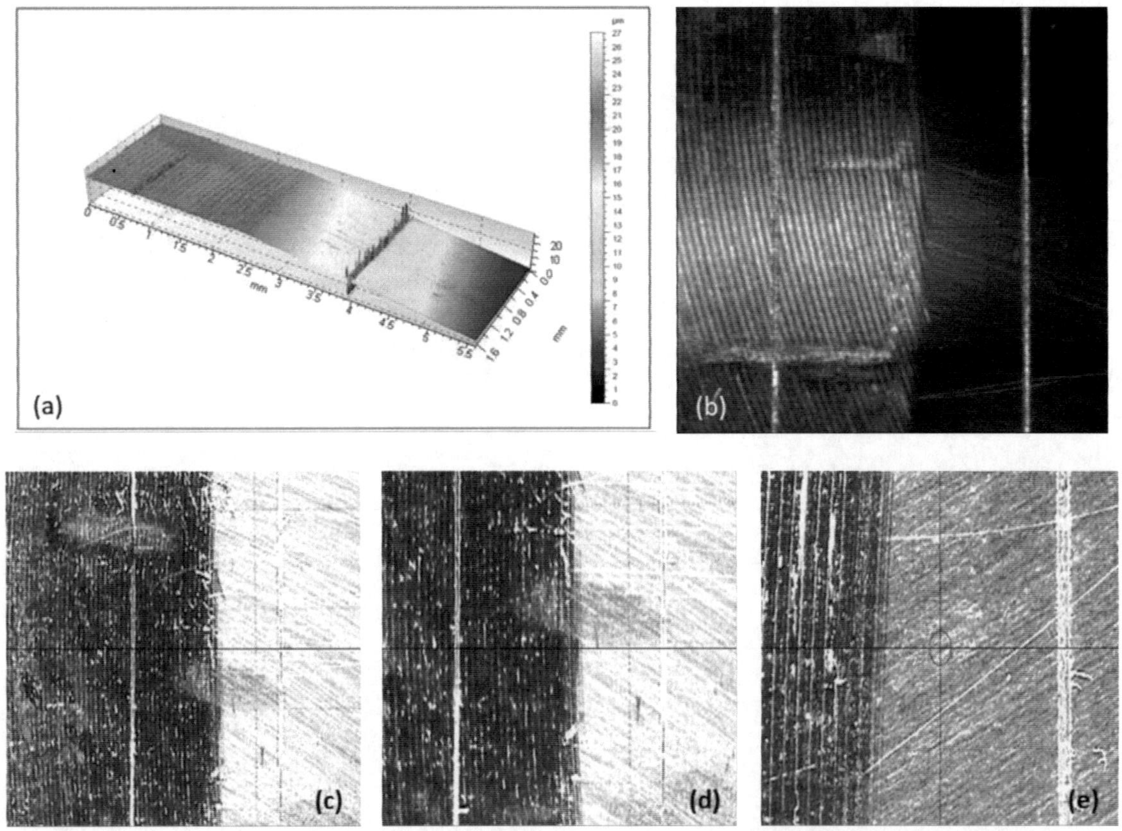

Figure 2. View of the surfaces (a) Talysurf CCI, sample No. 3 (b) O-Inspect sample No. 3 (c) MarVision, magnification x0.7 sample No. 5 (d) MarVision, magnification x1 sample No. 5 (e) MarVision, magnification x2 sample No. 3.

IOP Conf. Series: Journal of Physics: Conf. Series **1183** (2019) 012008 doi:10.1088/1742-6596/1183/1/012008

On the assumed probability P = 0.95, the experimental, relative method accuracy values were calculated using formulas 1 and 2. According to the Student's t-distribution, the parameter t value was determined. This value was equal 2.776.

$$w_h = \frac{h_i - h_{odn}}{h_{odn}} 100\% \quad (1)$$

$$DM = |\overline{w_h} \pm ts|_{max} \quad (2)$$

Where h_i - calculated value of the minimum chip thickness, h_{odn} - reference value of the minimum chip thickness, s - mean deviation.

The presented method allows to evaluate the possibility to employ the selected measuring devices to assess the value of h_{min} parameter. According to [18] the obtained values allow to quantitatively and qualitatively assess the relative method accuracy of evaluation the aforementioned parameter for selected devices.

Table 2. Ranges of relative method errors in surface texture measurement and the corresponding applications [18].

Measurement accuracy range (%)	Type of application
2÷5	Measurement of standards: roughness, waviness, form profiles
5÷15	Scientific research
10÷25	Measurement of surface texture under industrial condition

In order to investigate the possibility of application the selected measuring machines there were prepared five samples. The first research phase involved the assessment of the h_{min} value in directly manner. The value was calculated based on the measuring points from the optical profilometer and it was adopted as reference value. The prism draft angle and the length of the zone where the milling process was not regular were taken into account to assess this value. The minimum chip thickness was calculated using trigonometric function.

The analogous procedure of assessment the minimum chip thickness was used for all measuring devices. However, in order to precise evaluation the locus of zone of initiating of cutting process it was performed filtration process of the images of surface layer. In this research, two-dimensional wavelet transform (mother wavelet *bior6.8*) and thresholding method were applied to better visualize the characteristic cutting zones. The obtained signals are shown in figures 3-7, whereas the calculated values of the parameter h_{min} are presented in table 3.

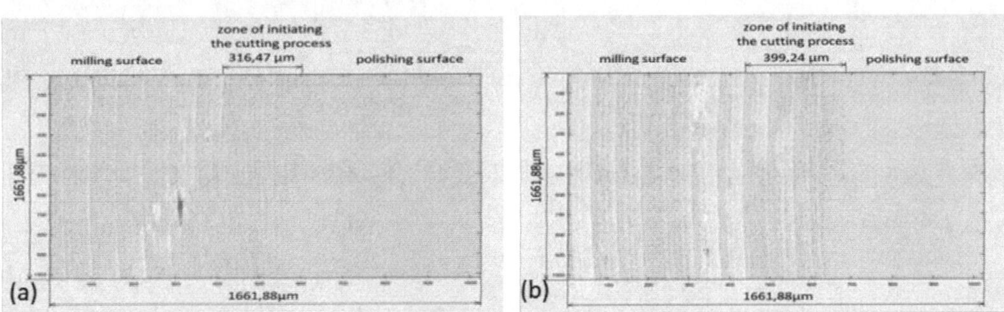

Figure 3. View of the wavelet decomposition coefficients on the surface, Talysurf CCI (a) sample No. 2 (b) sample No. 4

IOP Conf. Series: Journal of Physics: Conf. Series **1183** (2019) 012008 doi:10.1088/1742-6596/1183/1/012008

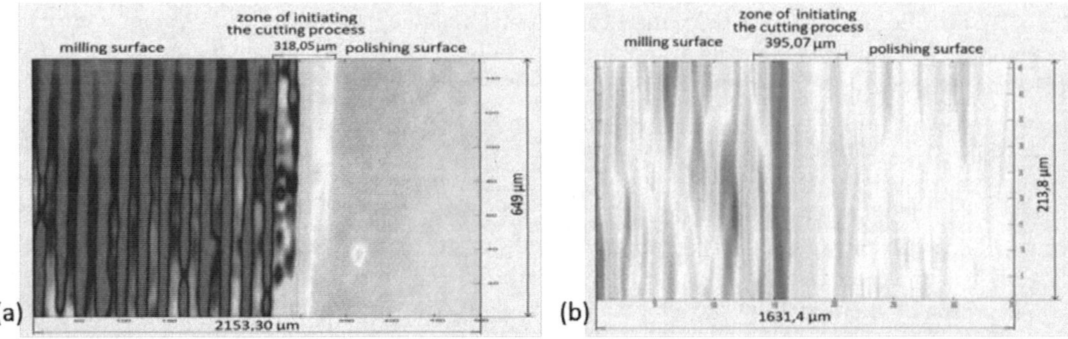

Figure 4. View of the wavelet decomposition coefficients on the surface, O-Inspect a) sample No. 3 b) sample No. 5

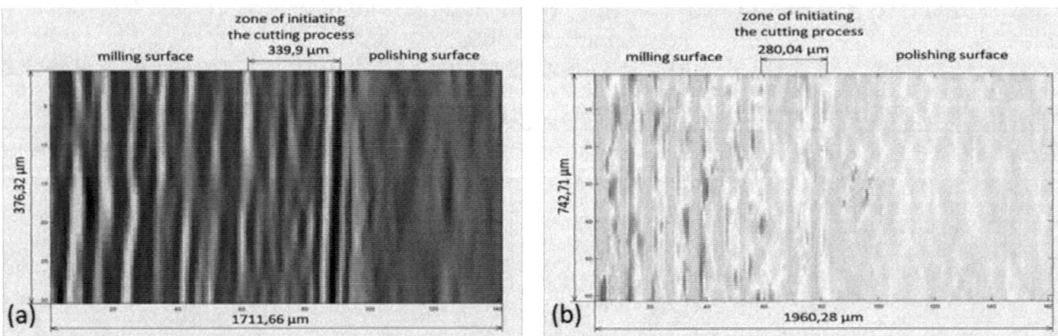

Figure 5. View of the wavelet decomposition coefficients on the surface, MarVision, magnification x0.7 a) sample No. 2 b) sample No. 3

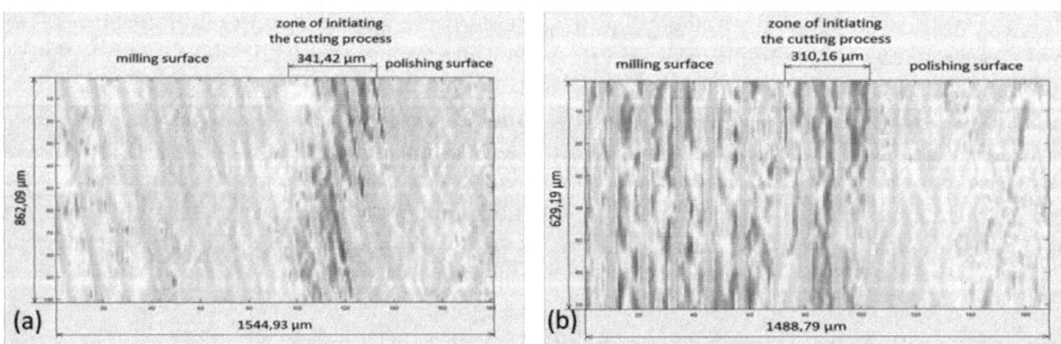

Figure 6. View of the wavelet decomposition coefficients on the surface, MarVision, magnification x1 a) sample No. 2 b) sample No. 3

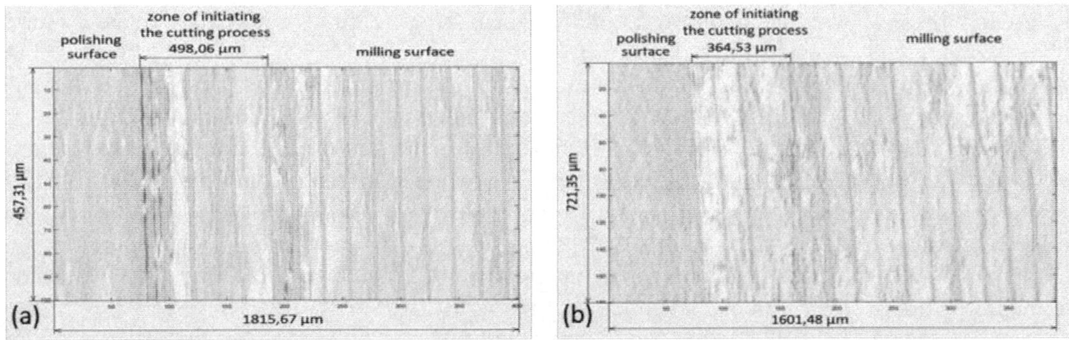

Figure 7. View of the wavelet decomposition coefficients on the surface, MarVision, magnification x2 a) sample No. 1 b) sample No. 4

Table 3. The values of the h_{min} parameter

No.	draft angle	Talysurf CCI (µm)		O-Inspect (µm)	MarVision (µm)		
		reference value	wavelet analysis		x0.7	x1	x2
1	0°33'59"	2.64	2.56	2.67	2.89	2.90	2.79
2	0°36'14"	2.23	2.09	2.41	2.25	2.26	2.38
3	0°37'23"	2.22	2.12	2.18	1.92	2.12	2.14
4	0°35'44"	2.43	2.60	2.50	2.59	2.50	2.38
5	0°30'54"	2.27	2.32	2.20	1.77	2.04	2.11

Analyzing the surface images obtained as a result of application two-dimensional wavelet transform, it can be noted that despite the performing the filtration process, in some cases there were not possible to clearly determine the cutting zones during the initiation of the cutting process. The evaluation of characteristic zones require significant skill from the person performing the analysis. However, the h_{min} parameter values presented in the above table for particular measuring machines were only slightly differ than reference values.

Formulas 1 and 2 were used in order to comprehensive evaluation of the possibility to apply the selected measuring machines to assess the minimum chip thickness. The DM parameter values which indicate the relative method accuracy are presented in table 4.

Table 4. Relative method accuracy, (%)

	Talysurf CCI	O-Inspect	MarVision		
			x0.7	x1	x2
DM	12.49	12.53	29.8	18.58	14.94

IOP Conf. Series: Journal of Physics: Conf. Series **1183** (2019) 012008 doi:10.1088/1742-6596/1183/1/012008

The research result shown that all above-mentioned measuring machines can be used to assess the value of the h_{min} parameter. The obtained values were in range of 12.49% to 29.8%. According to table 2, it can be concluded that the method which based on two-dimensional wavelet analysis can be used for scientific research. The most favourable result was obtained for the signals obtained from the optical profilometer (12.49%). This fact was probably related with the density of the measurement points and the exact surface mapping. A slightly higher value was obtained for the O-Inspect multisensor measuring machine (12.53%). The least favourable result was obtained for the MarVision workshop microscope. For the two magnification level, i.e., x0.7 (29.8%) and x1 (18.58%), the DM values were higher than 15%. It can also be stated that with the increasing the magnification level the obtained values decreased. For magnification x2, the relative method accuracy value was obtained on a satisfactory level of parameter value equal 14.94%.

4. Conclusion

The assessment of the minimum chip thickness value is key issue during choosing machining allowance for finishing process or the depth of cut, especially for machining for which depth of cut value is lowest than the value of radius of cutting insert. The evaluation of the h_{min} parameter value is a process which require experience to proper interpretation of characteristics zones of the impact of cutting insert on the workpiece during initiation of the cutting process. The application of devices with high vertical and horizontal resolution make possible to assess correctly the parameter value. The analysis of the possibility of application optical measuring machines which resolution in the particular axes were significantly lower than resolution of optical profilometer showed that application of wavelet transform allowed to proper detect characteristic cutting zones in images. For the particular measuring devices, the relative method accuracy values allow to conclude that the method based on two-dimensional wavelet analysis can be used for scientific research. The most favourable results were obtained for optical profilometer Talysurf CCI (12.49%) and the multisensory measuring machine O-Inspect (12.53%). The application of wavelet analysis enabled the correct identification of the characteristic zones during initiating a cutting process. Particular cutting zones were not visible on surface images prior to perform wavelet filtration.

The study also included the evaluation of the influence of the magnification of the MarVision MM320 workshop microscope on the possibility to correct identification of the above mentioned zones. According to table 4, with the increasing of the magnification, the DM parameter value decreased. Only for magnification x2 the parameter value indicated that the workshop microscope can be used to assess the value of the h_{min} parameter.

The research can be extended to evaluate the possibility of application two-dimensional wavelet transform to analyze images of surface layer obtained from other measuring machines or digital camera. Research can be carried out on a wider range of mother wavelet and milling process parameters for different types of cutting insert, in particularly for inserts which performed a specified number of work cycles.

References

[1] Adamczak S and Zmarzły P 2017 *Bull Pol Acad Sci-Te* **65** 541-51
[2] Kozior T and Kundera C 2017 *Procedia Engineer* **192** 463-8
[3] Cage J, Park S S and Freiheit T 2006 *Int. J. Mach. Tool. Manu.* **46**,313-32
[4] Blake P N and Scattergood R O 1990 *J Am Ceram Soc* **73** 949-57
[5] Malekian M, Mostofa M G, Park S S and Jun M B G 2012 *J Mater Process Tech* **212** 553-9
[6] Sai L, Bouzid W and Zghal A 2008 *J Mater Process Tech* **204** 213-20
[7] Nowakowski Ł, Miko E and Skrzyniarz M 2016 *Engineering Mechanics* (*Svratka*), pp 426-9
[8] Wieczorowski M 2009 *The use of topographic analysis of surface roughness measurements* (Poznań: Wydawnictwo Politechniki Poznańskiej) pp 156-8 (in Polish)
[9] Jiang X, Scott P and Whotehouse D 2008 *CIRP Ann.-Manuf. Techn.* **57** 555-8
[10] Zahouani H, Mezghani S, Vargiolu R and Dursapt M 2008 *Wear* **264** 480-5

IOP Conf. Series: Journal of Physics: Conf. Series **1183** (2019) 012008 doi:10.1088/1742-6596/1183/1/012008

[11] Abdul-Rahman H S, Jiang X and Scott P 2013 *Precis Eng* **37** 187-202
[12] Jiang X and Blunt L 2001 *Wear* **250** 217-21
[13] Stępień K and Makieła W 2013 *Metrol Meas Syst* **20** 139-50
[14] Dutta S, Pal S K and Sen R 2016 *Measurement* **77** 388-401
[15] Yang C, Liu P, Yin G, Jiang H and Li X 2016 *NDT&E INT* **83** 78-87
[16] Wang H X, Zong W J, Sun T and Liu Q 2010 *Appl Surf Sci* **256** 5061-8
[17] Talu S, Stępień K and Caglayan M O 2015 *Microsc Res Techniq* **78** 1026-31
[18] Adamczak S, Janusiewicz A, Makieła W and Stępień K 2011 *Metrol Meas Syst* **17** 35-46

IOP Conf. Series: Journal of Physics: Conf. Series **1183** (2019) 012009 doi:10.1088/1742-6596/1183/1/012009

Uncertainty evaluation for absolute flatness measurement on horizontally aligned fizeau interferometer

K Wirotrattanaphaphisan[1], J Buajarern[2] and S Butdee[1]
[1]Department of Production Engineering, Faculty of Engineering, King Mongkut's University of Technology North Bangkok, Bangkok 10800, Thailand.
[2]Dimensional Metrology Department, National Institute of Metrology (Thailand) Pathumthani, 12120, Thailand

E-mail: kibuya@gmail.com

Abstract. This paper describes and reports uncertainties associated with the absolute flatness measurement for an Ø 300 mm optical flat. Absolute flatness of 3 flats denoted as optical flat A, B and C was determined by conducting a three-flat test method. The measurements were carried out using a Fizeau interferometer. Sources of uncertainty that contributed to the measurement uncertainty were examined. The dominant contributions include off-center alignment during three-flat test measurement, temperature instability and deformation due to gravity. In our experiment, although the alignment was performed crucially, deviation up to 4 nm was observed from 6 sets of three-flat test. Since the optical flat was mounted using a sling type module, deformation due to clamping force can be neglected. Deformation due to gravitational force was investigated by comparing the contour map of the optical flat oriented at 0° and 180°. The maximum difference up to 99 nm was observed which is higher than accuracy of the reference flat ($\lambda/20$) itself.

Keywords: flatness, uncertainty, Fizeau interferometer

1. Introduction

The challenge of modern machining industries involves the achievement of high quality and high geometric and dimensional accuracy of work piece [1]. For better part functioning and assembly, part will be made up with good dimensional tolerance as well as with good geometric tolerance. Flatness is a surface form control [2]. Flatness is commonly used on planar surface capable of resting on matting planar surface without any significant rocking. Another important application of flat surface can be found in optics. The mechanism behind the formation of flatness as geometric tolerance is very dynamic, complicated, and process dependent. Several factors that influence the flatness in a CNC milling operations are spindle speed, feed rate, depth of cut and insert type for example.

A perfectly flat surface is defined as having all its elements in the same plane. The flatness tolerance zone consists of the distance between two parallel planes. In order to obtain a reliable assessment of flatness form, an appropriate extraction strategy for obtaining a representative set of points on the work piece is required. Flatness can be measured by various techniques which can be divided into 2 groups, contact type and non-contact type. The contact method can be conducted using dial indicator, height gauge and coordinate measuring machine (CMM) for instance. However neither of these methods can measure flatness more accurately than 2.5 μm [3-7]. Non-contact method is commonly used with lapped parts which is based on the reflection and interference of monochromatic light [8]. A monochromatic

IOP Conf. Series: Journal of Physics: Conf. Series **1183** (2019) 012009 doi:10.1088/1742-6596/1183/1/012009

light source and a reference flat are needed for this measuring method. The reference flat, a piece of transparent glass that has itself been lapped and polished on one or both sides, is placed on the lapped surface and plays role as the reference standard. The monochromatic light is then shone down through the glass. The light will pass through glass and reflect off the workpiece. As the light reflects in the gap between the workpiece and the polished surface of the reference flat the light will interfere with itself creating bright and dark fringes. Each fringe represents a change of one half wavelength in the gap width between the reference flat and the workpiece. The light bands display a contour map of the workpiece's surface and can be readily interpreted for flatness value [9]. The more advanced flatness measuring instrument with the highest accuracy is by using flatness interferometer. Fizeau interferometer is equipped with the He-Ne laser generating a single wavelength illumination and phase shifter. Phases of the interference pattern were determined from phase shifting algorithm and converted to distance between reference flat and the workpiece. The difference between the maximum point and the minimum point is referred to as flatness. The interferometric method of flatness measurement is a form of length measurement in which distances variation between a test surface and a reference surface is measured at many locations simultaneously. The phase differences between the two wavefronts (reflected and transmitted) result in fringe pattern and that is a direct indication of the form of the component being tested with respect to the reference flat [10-11].

Considering that flatness of the reference surface is better than the test surface's, thus all the differences are counted for the test surface. Therefore, the accuracy of measurement is directly limited by the quality of the reference flat.

In usual flatness test with the reference surface if errors of test surface are similar to those of reference surface, then it is impossible to identify which flat contributes the error. Absolute flatness measurement is an interferometric method to obtain the actual flatness of an optical surface independent of the flatness of reference flat used. In this method, 3 flats in a series of configurations are measured [12-13]. Then the flats are rotated and replaced during measurements in a prescribed order to overcome the error source from the reference flat. The absolute testing is actually beneficial, when we use a tested flat as a reference surface during measurements. The data obtained is extremely accurate and it allows high-accuracy characterization of any optical surfaces.

The uncertainty of measurement is associated with the result of a measurement that characterizes the dispersion of the values that could reasonably be attributed to the measurand [14]. Toshiyuki et al. demonstrated uncertainty evaluation for three-flat test measurement performed on the vertically aligned Fizeau interferometer. Sources of uncertainty are repeatability, reproducibility, temperature effect, deformation due to gravity and uncertainty of algorithm in three-flat test. The main sources of measurement uncertainty are the environment and gravitational force.

When measuring flatness of optics, surface of an unknown is measured comparatively to the reference surface. In order to determine the absolute flatness error of the unknown face, three-flat test method is need to be conducted. There has been a research reported the uncertainty estimation of the flatness measurement on the vertically aligned Fizeau interferometer. In order to increase accuracy of flatness measurement on horizontally aligned Fizeau Interferometer, uncertainty contribution is examined here.

In this paper, we explored the amount of deformation of the optics mounted horizontally. The three-flat test measurement of optical flat, 300 mm in diameter, on the horizontally aligned Fizeau interferometer was conducted. The optical flat were mounted on the sling type mounting in order to avoid deformation due to clamping force. The uncertainty contributed from environmental effect and deformation due to gravitational force were investigated.

2. Experiment setup

2.1. Instrument

The flatness interferometer used is manufactured by 4D Technology. This system can measure flatness of workpiece diameter up to 300 mm. The system equipped with the stabilized He-Ne laser, illuminating light at 632.8 nm, and phase shifter. Interference fringes were captured by the CCD camera with

IOP Conf. Series: Journal of Physics: Conf. Series **1183** (2019) 012009 doi:10.1088/1742-6596/1183/1/012009

resolution of 1200 pixel × 1200 pixel. The system were placed on the anti-vibration table. Figure 1 illustrates schematic diagram of the horizontally aligned flatness interferometer system.

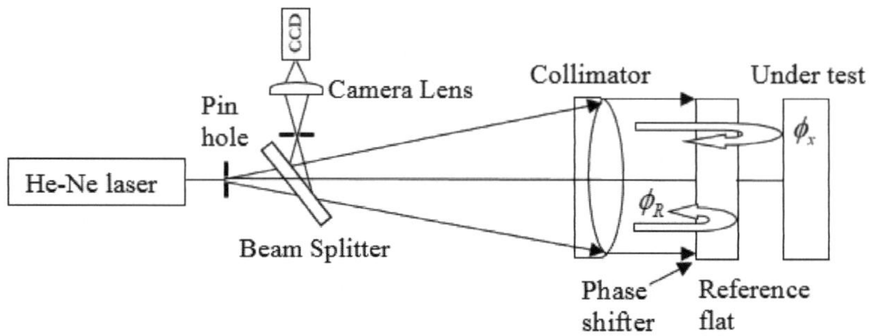

Figure 1. Schematic diagram of the flatness interferometer.

Phases (ϕ) of the interference fringes were determined using 9-steps phase shifting algorithm designed by Phillion D.W. [15]. Phase difference was then converted to distance in unit of nanometer creating a surface topography of the optic under-tested. Flatness value was determined from the difference between the maximum point (peak) and the minimum point (valley). Figure 2 shows optical flat rested on a sling and mounted on the tip/tilt stage. The sling is made of aluminum, 610 mm in length and 9.4 mm in width.

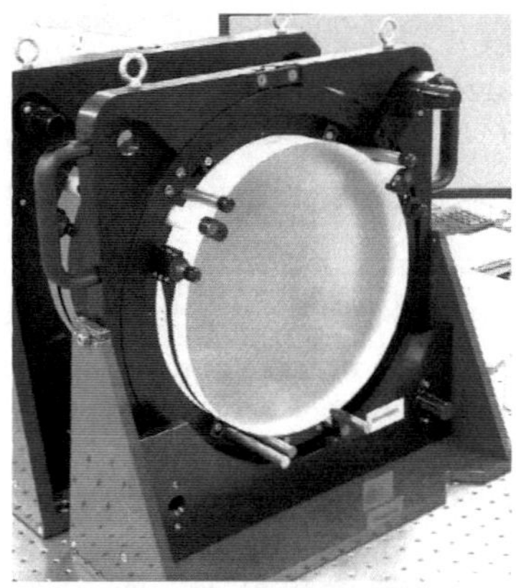

Figure 2. Optical mounting system.

Optical flat A, B and C are 300 mm in diameter with thickness of 50.8 mm. Their weight are approximately 10 kg. The experiments were employed in the environmental controlled laboratory. The environmental condition is $20 \pm 1\,^{\circ}C$ for temperature and 50 ± 10 % RH for relative humidity.

2.2. Three-flat test method

Three-flat test method helps eliminates form errors of the reference flat. It is ideal for calibrating reference flats because it gives an absolute calibration. There are various methods for absolute flat testing. Also, several algorithms have been developed to determine absolute surface error [16-21].

The experimental procedure consists of 4 sets of measurement [22-23]. Figure 3 demonstrates configurations of optical flat setup for four measurement require for the three-flat test method. The optical flat to be measured in are labeled as A, B and C. The first set denoted as measurement I, consists of flat B acting as transmission flat and flat A as a reference flat. The second set denoted as measurement II, consists of flat B acting as transmission flat and flat A is rotate 90 degree as a reference flat. The third

IOP Conf. Series: Journal of Physics: Conf. Series **1183** (2019) 012009 doi:10.1088/1742-6596/1183/1/012009

set denoted as measurement III, consists of flat C acting as transmission flat and flat A as a reference flat. The last set denoted as measurement IV, consists of flat C acting as transmission flat and flat B as a reference flat.

Absolute flatness testing is an interferometric method for obtaining the actual surface error of flat object, independent of flatness error of the reference flat used. This method, one measures three flats in a series of configurations, rotating and replacing flats between measurements in a prescribed order to eliminate the reference surface's flatness as an error source. The analysis of measurement results was done by using even and odd functions. The function of even-odd, odd-even, even-even, and odd-odd are illustrated as in (1) - (3). With rotation is required in the test method, horizontally mounted flat experience gravitational effect differently between each measurement series. It should be noted that when the measurement sequence as shown in figure 3, the optical flat was measured with an upward and a downward posture. The gravitational effect is needed to be examined as it directly affects accuracy of the absolute measurement result.

$$A = A_{ee} + A_{oe} + A_{eo} + A_{oo,\ 2odd\theta}\cdots \quad (1)$$
$$B = B_{ee} + B_{oe} + B_{eo} + B_{oo,\ 2odd\theta}\cdots \quad (2)$$
$$C = C_{ee} + C_{oe} + C_{eo} + C_{oo,\ 2odd\theta}\cdots \quad (3)$$

The even-odd, and odd-even function of absolute flatness can be analysis by three-flat test method using 2oddθ from four configurations [12].

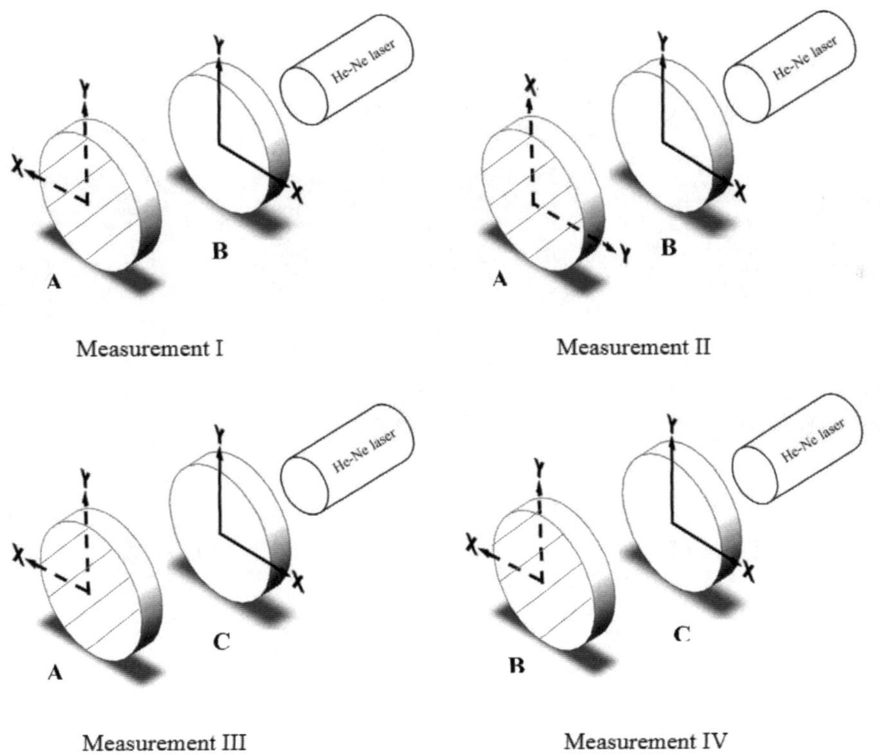

Figure 3. Configurations of four required for three-flat test method.

IOP Conf. Series: Journal of Physics: Conf. Series **1183** (2019) 012009 doi:10.1088/1742-6596/1183/1/012009

3. Results and discussion

3.1. Three-flat test measurement

Figure 4 illustrates surface contour maps obtained from measurement I, II, III and IV. Each measurement is obtained from averaging 20 repeated measurements. The three-flat test measurements were reproduced 6 times.

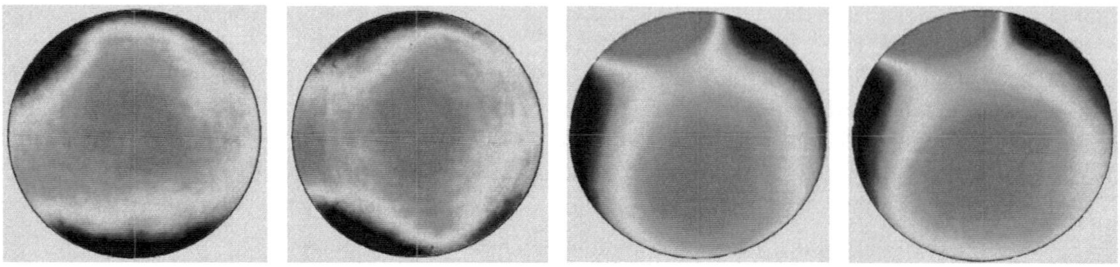

(a) (b) (c) (d)

Figure 4. The measurement results obtained from (a) measurement I, (b) measurement II, (c) measurement III and (d) measurement IV.

It should be emphasized that such results are not absolute form error of optical flat but combined error of 3 optical flats. The absolute form error of optical flat A, B and C were calculate according to as in (1) - (3) yielding surface contour of optical flat A, B and C as shown in figure 5.

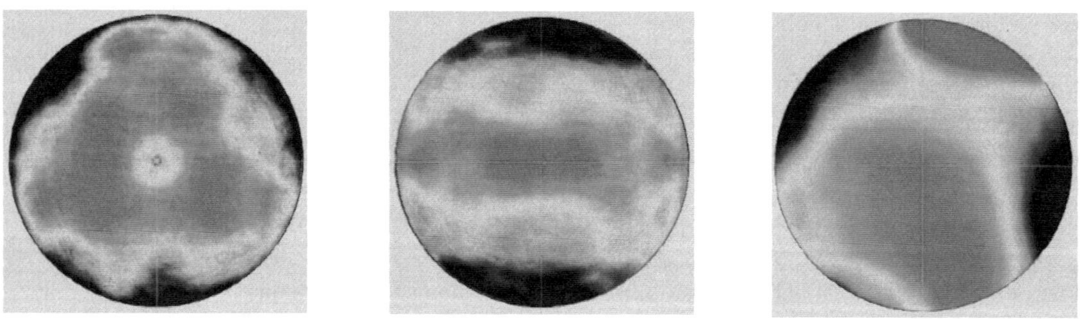

(a) (b) (c)

Figure 5. The contour map of optical flat (a) A, (b) B and (c) C.

Table 1. PV value of optical flat A, B and C

Flat	PV (nm)	σ_{n-1} (nm)
A	61.92	3.06
B	71.61	4.24
C	393.01	7.11

The averaged peak-to-valley (PV) and standard deviation of optical flat A, B and C obtained from 6 measurements were determined and were summarized in table 1.

IOP Conf. Series: Journal of Physics: Conf. Series **1183** (2019) 012009 doi:10.1088/1742-6596/1183/1/012009

3.2. Uncertainty evaluation

The uncertainty was evaluated in accordance with the JCGM 100:2008 (Evaluation of measurement data - Guide to the expression of uncertainty in measurement) [24]. The measurement uncertainty can be determined by combining all sources of uncertainty as deserved as in (4)

$$u_c^{\,2}\ (x) = c_1^{\,2}u^2(x_1) + c_1^{\,2}u^2(x_2) + \ldots + c_n^{\,2}(x_n) \tag{4}$$

where $u(x)$ is combined standard uncertainty, $u(x_i)$ is standard uncertainty of component i and c_i is sensitivity coefficient of component i. Flatness of optical flat was calculated from the difference between the maximum height $(x_{\max})$ and the minimum height $(x_{\min})$ of the flat's topography.

$$PV = x_{\max} - x_{\min} \tag{5}$$

The surface height can be determined from phase (ϕ) as in (6) when λ is wavelength of light source.

$$x = \frac{\phi}{4\pi}\lambda \tag{6}$$

Phase value was analyzed from 9-step phase shifting algorithm according to as in (7) when I_i is intensity of light detected by the CCD camera and i is the step order [25].

$$\phi = \tan^{-1}\left(\frac{-4I_2 + 12I_4 - 12I_6 + 4I_8}{I_1 - 8I_3 + 14I_5 - 8I_7 + I_9}\right) \tag{7}$$

From as in (5) to as in (7), uncertainty of flatness can be written as following

$$\begin{aligned} u^2(PV) &= u^2(x_{\max}) + u^2(x_{\min}) + u^2(x_{3-flat}) \\ &= 2\left(c^2{}_x u^2(x)\right) + \left(c^2{}_{x_{3-flat}} u^2{}_{x_{3-flat}}\right) \end{aligned} \tag{8}$$

where $u(x_{3-flat})$ is the standard uncertainty due to absolute calibration using 3-flat test method. From as in (4), and as in (6) can be rewritten as

$$u^2(x) = \left(\frac{\lambda}{4\pi}\right)^2 u^2(\phi) + \left(\frac{\phi}{4\pi}\right)^2 u^2(\lambda) \tag{9}$$

Thus, as in (8) can be described as

$$u^2(PV) = 2\left[\left(\frac{\lambda}{4\pi}\right)^2 u^2(\phi) + \left(\frac{\phi}{4\pi}\right)^2 u^2(\lambda)\right] + u^2(x_{3-flat}) \tag{10}$$

3.2.1. Uncertainty due to phase determination, $u(\phi)$

Phase is the position of point on waveform. The uncertainty of phase is caused by several factors including nonlinearity of piezoelectric transducer (PZT), background noise, signal drift, sensitivity of

IOP Conf. Series: Journal of Physics: Conf. Series **1183** (2019) 012009 doi:10.1088/1742-6596/1183/1/012009

the detector and vibration Phillion D.W. suggested that by analysing phase using as in (7), phase error became insensitive to the detector's sensitivity and vibration. Phase error is limited to calculation as in (11).

$$\left\{(\delta\theta)^2\right\}^{\frac{1}{2}} / 2\pi = 2.261\times10^{-4}\left(1-0.04\sin^2\theta\right)^{\frac{1}{2}}, wave \quad (11)$$

The standard uncertainty of phase determination is considered as rectangular distribution and its value is equal to 0.00082.

3.2.2. Uncertainty due to accuracy of wavelength, $u(\lambda)$

Wavelength of laser plays a crucial role in converting. Phase value to height value. Hence, accuracy of height value strongly depends on accuracy of the laser's wavelength and decimal place of the wavelength value used in calculation. Both errors are considered as rectangular distribution and the standard uncertainty of 0.00577 nm was obtained.

3.2.3. Uncertainty in 3-flat test measurement, $u(x_{3-flat})$

Upon conducting 3-flat test method, the measurement was performed at 4 different configurations as illustrated in figure 3. There are various factors that affect accuracy of this absolute calibration. There are temperature fluctuation $(u(x_{3-flat-1}))$, centering error $(u(x_{3-flat-2}))$ and deformation due to gravity $(u(x_{3-flat-3}))$.

For each measurement, the optical flat were left to be stabilized with the setup and temperature for at least 8 hours. Thus, one set of 3-flat test calibration required approximately 4 days to complete. Therefore, effect of temperature fluctuation to flatness measurement was investigated.

Figure 6 illustrates height variation of the optical flat C at position X1 (-147.75 mm, 0 mm), X2 (-134.53 mm, 0 mm), Y1 (0 mm, -139.29 mm) and Y2 (0 mm, -99.64 mm). The variation was examined after 8 hours stabilization. The height variation of ± 32 nm and the normal distribution with maximum experimental standard deviation of 13.55 nm were observed.

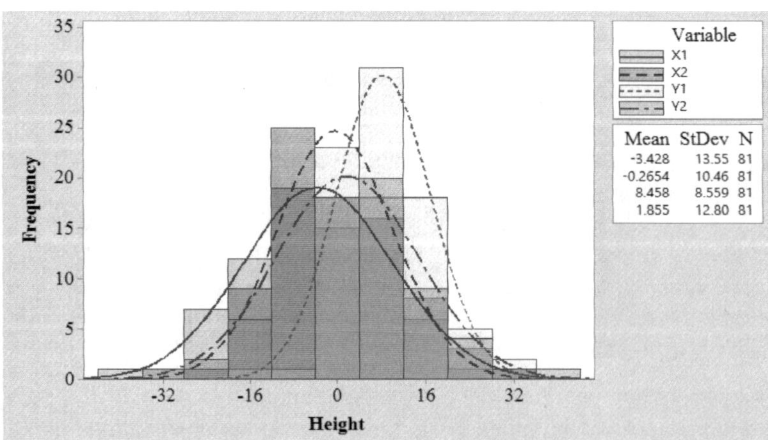

Figure 6. Variation in calculated distance at four different positions of reference flat C.

Centering is also strongly effect accuracy of 3-flat test analysis. Alignment with out of center by only 5 nm can cause error in absolute calibration. We examined alignment error by reproducing 3-flat test measurement from 6 set of averaged height value and its experimental standard deviation of optical flat A, B and C at position (75 mm, 75 mm), (-75 mm, 75 mm), (-75 mm, -75 mm) and (75 mm, -75 mm) denoted as position 1-4, respectively, are summarized in table 2. The maximum standard deviation of 5.28 nm was observed.

Table 2. Height of optical flat at position 1, 2, 3 and 4

Flat	Position	$\bar{X}$ (nm)	σ_{n-1} (nm)
A	1	2.04	3.84
	2	-0.63	2.00
	3	7.20	4.61
	4	4.71	2.97
B	1	-2.94	2.15
	2	1.80	4.05
	3	-3.60	1.84
	4	0.29	2.38
C	1	-28.22	3.68
	2	-54.26	1.59
	3	16.68	5.28
	4	15.15	3.54

3-flat test method requires rotation of the optical flat. Unlike vertically aligned flatness interferometer, the deformation due to gravity is non uniform throughout the measurement for the horizontally aligned system. Deformation due to gravitational force was investigated by comparing the profile of optical flat C along y-axis with those of the optical flat C after 180° rotation and the deviation. The maximum difference of 99 nm was observed. Figure 7 illustrates result the measured of cross section profile along y-axis of the reference flat C aligned at 0° and 180°. According in to the guideline [26], the deformation of optical flat with the theoretically perfect edge support was calculated to be approximately 90 nm which is in good agreement with our experimental result.

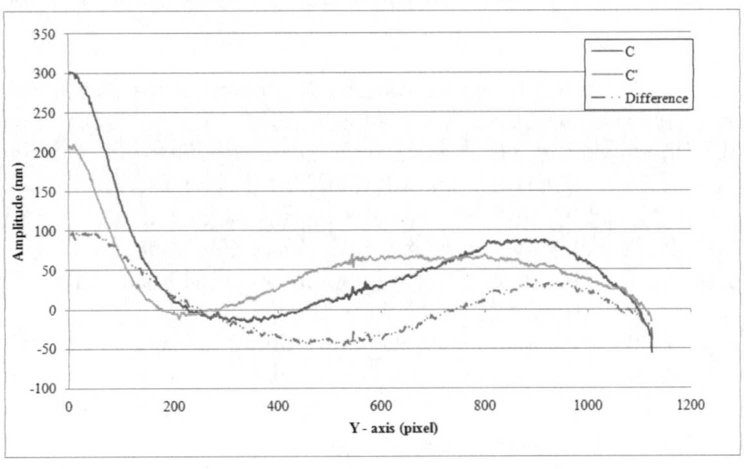

Figure 7. Cross section profile along y-axis of the reference flat C aligned at 0° and 180° and the difference between the two profiles.

As the result, the standard uncertainty in 3-flat test measurement became 57.22 nm. Uncertainty budget of flatness value is summarized in table 3. The expanded uncertainty of PV value at the confidence level of approximately 95% is 115 nm.

Table 3. Uncertainty budget

Symbol	Value	Unit	Standard uncertainty	Probability distribution	Sensitivity coefficient	Uncertainty contribution	Unit
$u(\lambda_1)$	6.33E-06	nm	3.65E-06	Rectangular	0.056	2.06E-07	nm
$u(\lambda_2)$	0.01	nm	5.77E-03	Rectangular	0.056	3.25E-04	nm
$u(\phi)$	1.42E-03	rad	8.20E-04	Rectangular	50.357	4.13E-02	nm
$\boldsymbol{u(x_i)}$				**Normal**		**4.13E-02**	**nm**
$u(x_{3\text{-}flat1})$	13.55	nm	1.51	Normal	1	1.51E+00	nm
$u(x_{3\text{-}flat2})$	5.28	nm	2.16	Normal	1	2.16E+00	nm
$u(x_{3\text{-}flat3})$	99	nm	57.16	Rectangular	1	5.72E+01	nm
$\boldsymbol{u(x_{3\text{-}flat})}$				**Normal**		**5.72E+01**	**nm**
$\boldsymbol{u(PV)}$				**Normal**		57.22	nm
$\boldsymbol{U_{95\%}(PV)}$				**Normal**		115	nm

4. Conclusions

We found that the major source of uncertainty is the environmental fluctuation and deformation due to the gravitational force. Optical flats were absolute calibrated using 3-flat test method performing on the horizontally aligned Fizeau interferometer. Absolute flatness error of optical flat A, B and C were deformed and the PV value are 61.92 nm, 71.61 nm and 393.01 nm respectively. The measurement uncertainty was evaluated and the expanded uncertainty of 115 nm at 95% confidence was obtained. The major source of uncertainty is deformation of the optical flat due to gravitational force. In order to reduce the measurement uncertainty to be within accuracy of the reference flat ($\lambda/20$), deformation due to gravity shall be corrected.

Acknowledgments

The authors would like to express their appreciation to the National Institute of Metrology (Thailand) for allowing usage of instrument and infrastructure. Rajamangala University of Technology Lanna is grateful for scholarship of KW.

References

[1] Rashid M A and Lani M R 2010 Surface roughness prediction for CNC milling process using artificial neural network *Proc.Eng.* **3**.

[2] Henzold G 2006 *Handbook on the Geometrical Dimensioning and Tolerancing for Design, Manufacturing and Inspection* (Burlington Jordan-Hill: Elsevier) p 4.

[3] Modi B S, George P M and Sheth S 2013 Study and investigate effect of cutting parameters on flatness for dual plate check valve *Int. Conf. Adv. Comp. Commun. Tech.*

[4] Lakota S, Gorog 2011 A flatness measurement by multi-point methods and by scanning methods *J. interdisciplinary research. Phys.* **2** 2.

[5] Joshi A and Kothiyal P 2013 Investigating effect of machining parameters of CNC milling on surface finish by taguchi method *Int. J. on Theoretical and Appl. Research Mech. Eng.* **1**.

[6] Pai D, Rao S and Shetty R 2010 Application of response surface methodology on surface roughness in grinding of aerospace materials *J. Eng. Appl. Sci.* **5** 6.

[7] Quick guide to precision measuring instruments Mitutoyo Corporation Kanagawa.

[8] Shashank S, Bhamble Surface finishing *Adva. Manu. Tech.* Shegaon.

[9] JIS B 7430:1977 Optical flat Japanese Industrial Standard. (Edition).

[10] Takatsuji T, Osawa S, Kuriyama Y and Kurosawa T 2003 Stability of the reference flat used in fizeau interferometer and its contribution to measurement uncertainty *Proc. of SPIE* Vol **5190.**

[11] Xu C, Chen L and Yin J 2009 Method for absolute flatness measurement of optical surfaces *Appl. Opt.* **48** 13.

IOP Conf. Series: Journal of Physics: Conf. Series **1183** (2019) 012009 doi:10.1088/1742-6596/1183/1/012009

[12] [Online]. Available: www.veeco.com/pdfs/appnotes/an506-absolute_05083_352.pdf.
[13] Reeve C P 1975 The calibration of an optical flat by interferometric comparison to a master optical flat *Final Report National Bureau of Standards Washington DC.*
[14] ISO: 3534-1 General statistical terms and terms used in probability: Statistics-Vocabulary and symbols-Part 1, 2006.
[15] Phillion D W 1997 General methods for generating phase-shifting interferometry algorithms *Appl. Opt.* **36** 31.
[16] Schulz G 1967 Ein interferenzverfahren zur absolute ebenheitsprufung langs beliebiger zentralschnitte *J. of Modern Opt.* **14** 4.
[17] Schulz G and Schwider J 1976 Interferometric testing smooth surfaces *Progress in Optics* **13.**
[18] Schulz G and Grzanna J 1992 Absolute flatness testing by the rotation method with optimal measuring error compensation *Appl. Opt.* **31** 19.
[19] Grzanna J and Schulz G 1990 Absolute testing of flatness standards at square-grid points *Opt. Commun.* **77** 2-3.
[20] Fritz B S 1984 Absolute calibration of an optical flat *Opt. Eng.* **23** 4.
[21] Ai C and Wyant J C 1993 Absolute testing of flats by using even and odd functions *Appl. Opt.* **32** 25.
[22] Moona G, Sharma R, Kiran U and Chaudhary K P 2014 Evaluation of Measurement Uncertainty for Absolute Flatness Measurement by Using Fizeau Interferometer with Phase-Shifting Capability *J. Met. Soci. Indi.* **29** 4.
[23] Takatsuji T, Ueki N, Hibino K, Osawa S and Kurosawa T 2001 Japanese ultimate flatness interferometer (FUJI) and its preliminary experiment *Proc. SPIE Int. Soci. Opt. Eng.* 4401.
[24] JCGM 100:2008 GUM 1995 with minor corrections Evaluation of measurement data - Guide to the expression of uncertainty in measurement.
[25] Creath K 2000 Choosing a Phase Measurement Algorithm for Measurement of Coated LIGO *Int. Proc. SPIE* **4101.**
[26] Bryan J K 1978 A user's guide to designing and mounting lenses and mirrors (No. UCRL-52411). *University of California Lawrence Livermore Laboratory*

IOP Conf. Series: Journal of Physics: Conf. Series **1183** (2019) 012010 doi:10.1088/1742-6596/1183/1/012010

Fast Measurement Method of Gear Parameters Based on Convexity Defect

Chen Guo-hua, Liu Xiao-long, Zhang Ai-jun and Wang Jun-yi
Beijing University of Chemical Technology

E-mail: chengh@mail.buct.edu.cn

Abstract: Gear parameters measurement based on machine vision is an important part in the field of intelligent manufacturing. Existing methods, which are all based on gear center, may cause accumulative error and have the weakness of large computation and low efficiency due to the repeated use of all contour points. In view of these problems, a new measurement method of gear parameters based on convexity defect is presented. Firstly, a gear contour is extracted from the gear image. Secondly, convex hull and defects of the gear contour are computed using image processing approach, then the number of the gear teeth and key points on the dedendum circle and addendum circle are determined, respectively. Finally, each radius of the dedendum circle and addendum circle is obtained separately by means of circle fitting with the corresponding key points. In this method, the key points are calculated by using the properties of convex defects, avoiding the calculation of the gear center and greatly increasing the computing speed by using a limited number of key points, precisely, two times the number of gear teeth. Compared with the traditional methods, the experimental results show that the speed of this method is increased by 1.5 times and it can meet the real-time requirement.

Keywords: Gear, Parameters Measurement, Convexity Defect

1. Introduction

Gear is one of the most important parts in mechanical transmission, whose quality and accuracy affect the performance and service life of a machine. The gear's fast measurement, therefore, is necessary in real gear inspection. However, it is very difficult to perform precise measurements fleetly due to complicated shapes and varied features of a gear (Yin *et al* 2013). Gear parameters are usually artificially measured in conventional measurement methods, which is inefficient and prone to subjective misreading. With the rapid development of computer vision technology, visual measurement has become a research hotspot in recent years for the characteristics of real time, non-touching and high automation (Ma and Zhang 1998, Lin *et al* 2015, Klancnik *et al* 2015, Simunovic *et al* 2016, Wang and Mottershead 2016, Ge *et al* 2017).

According to the requirements of the gear's on-line detection and fast manufacturing, researchers have applied computer vision to gear parameters measurement. Zhang *et al* (2004) studied a non-contact measuring method for spur gears based on digital image processing technology, in which the gear geometric center is obtained by center-of-gravity method and the teeth number is calculated by drawing an auxiliary circle. As the method of Li and Liu (2009), the edge contour of a gear is firstly extracted using edge detection, then the center of the gear is computed by means of random Hough transform. Chen and Li (2010) use a similar method to Li and Liu's method. Wang (2011) proposed a method based on mathematical morphology for gear center localization and gear teeth number

IOP Conf. Series: Journal of Physics: Conf. Series **1183** (2019) 012010 doi:10.1088/1742-6596/1183/1/012010

counting, in which the gear center is obtained through fitting a circle with the least squares method, and edge detection is uniquely not needed.

In almost all the existing methods, the gear center is the basis to calculate other geometric parameters. After obtaining the gear contour, various methods such as the least squares method and the random Hough transform were applied to gain the gear center. But these methods all have their own shortcomings more or less. The least squares method itself has calculation error which will lead to positioning error of the center, pointed by Zhang *et al* (2004). In the process of extracting the circle and calculating its center by the random Hough transform, the choice of data may cause positioning error, as Li and Liu (2009) said in their paper. The positioning error of the center can subsequently result in the error of the other parameters of the gear, which will affect the accuracy of the measurement.

Moreover, in the existing methods, all the points on the contour will be utilized almost in each stage of calculations. Hence the amount of calculation is relatively large. By analyzing the gear shape, we found that convex hull and convexity defects may become the key to solve the problem, which is often used for gesture recognition (Mykyta *et al* 2014).

In this paper, a fast measurement method to calculate the gear parameters by using a finite number of key points on the contour was proposed, which greatly reduces the computational complexity and significantly improves the detection efficiency. Particularly, the gear center does not need to be positioned because this method does not require it to be the basis for calculating other geometric parameters. Based on the extracted key points, which is shown in figure 1 as the blue and pink circles, it is easy to find dedendum circle and addendum circle, the two bigger yellow circles below, by fitting the key points. And then the radii are not difficult to calculate.

Figure 1. Key points of gear contour and fitted circles based them

2. Acquisition of convex hull and defect of gear contour

2.1. Analyses of spur gear shape

From the appearance point of view, the teeth of spur gear are periodic with the nature of uniform distribution. Gear tooth and tooth space appear alternately on the outer edge of the gear and both are

IOP Conf. Series: Journal of Physics: Conf. Series **1183** (2019) 012010 doi:10.1088/1742-6596/1183/1/012010

evenly distributed around the center of the gear. Obviously, the number of gear teeth equals to the number of teeth space. To calculate the number of gear teeth, radius of the dedendum circle and addendum circle, all needed is a point on the dedendum circle and addendum circle respectively and the gear center, ideally. The point on the dedendum circle, which is called P1 in this paper, is the lowest point in the tooth space and the point on the addendum circle, which is called P2, is the highest point of on the gear tooth. Points like P1 and P2 are called key points.

$$\begin{Bmatrix} \sum P1 = \sum p2 = z \\ \sum Keypoint = \sum p1 + \sum p2 = 2z \end{Bmatrix} \tag{1}$$

Based on the above considerations, the problem of how to get the gear parameters converts to how to get the key points of the gear.

This paper presents a method, which extracts the key points from the convex hull and convexity defect of the gear contour and after that one can calculate the gear parameters directly and simply. Compared with existing methods, this method does not need to determine the gear center, use contour extraction to instead the edge detection and is not necessary to calculate all the pixels on the contour, reduce the amount of calculation greatly, significantly improve the detection efficiency.

2.2. Extraction of the gear contour

As shown below in figure 2, image preprocessing such as grayscale transformation and threshold segmentation should be performed firstly before extracting the gear contour. In this paper, the contour extraction refers to acquiring a set of points with coordinate information, which constitutes the contour. Subsequent calculations are based on these points.

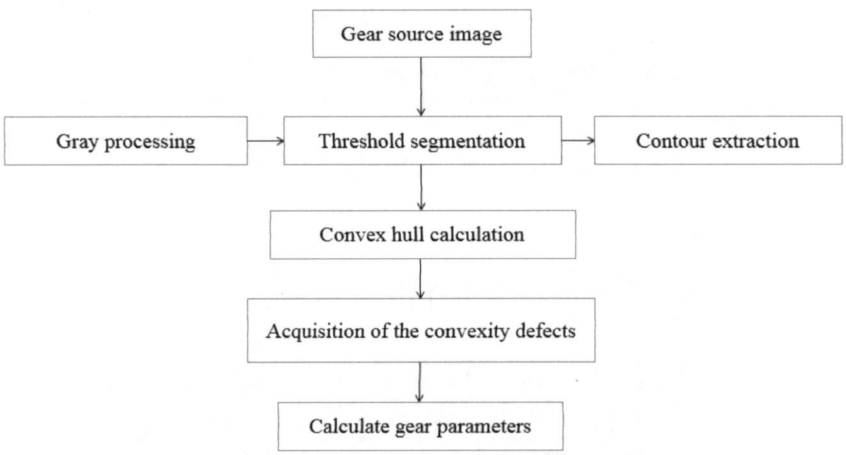

Figure 2. The flow chart of measurement process

A border following method (Suzuki and Be 1985) instead of edge detection like Canny's method that is commonly used at present is applied to get the gear contour. The advantage of using the border following method is that it can easily handle gear images with partially highlight areas (as illustrated in figure 3), while using the edge detection will obtain a lot of redundant information that must be removed by additional treatments.

IOP Conf. Series: Journal of Physics: Conf. Series **1183** (2019) 012010 doi:10.1088/1742-6596/1183/1/012010

Figure 3. (a) Original gear image (b) Binary image.

Figure 4(a) depicts the contour that was achieved from figure 3(b) by the border following method, and figure 4(b) shows the edges acquired using the Canny's method. Obviously, the latter contains much redundancy and needs additional treatments.

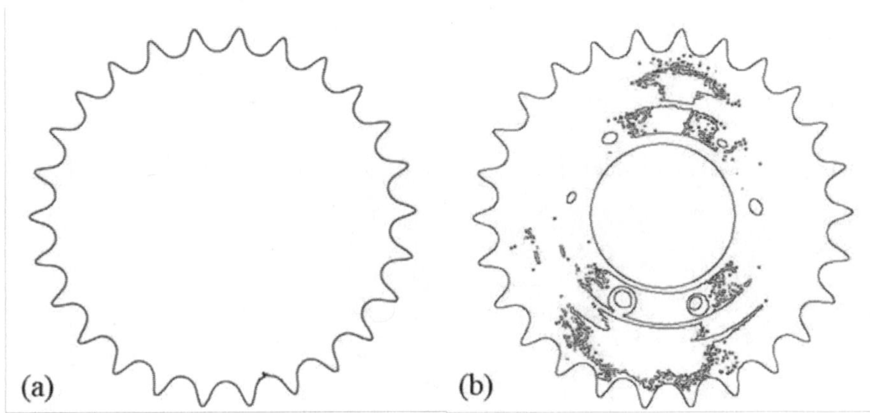

Figure 4. (a) Gear contour (b) Gear edge.

Another difference between the two methods is the form of outputs. The border following method can directly output the desired points composing the contour, while edge detection methods will give an image of gear edges, and the edge images should be further processed in order to obtain the points.

Figure 5 shows a common example in which the image does not contain superfluous information like highlight area as in figure 3. Figure5 (b) displays the binary image preprocessed from figure 5(a), and figure 5(c) depicts the gear contour extracted by border following method.

IOP Conf. Series: Journal of Physics: Conf. Series **1183** (2019) 012010 doi:10.1088/1742-6596/1183/1/012010

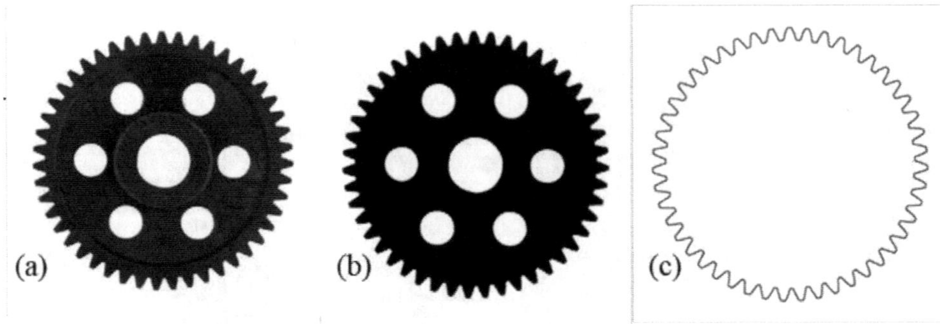

Figure 5. (a) Original gear image (b) Binary image (c) Gear contour image.

2.3. Convex hull acquisition

In order to extract the key points on the top circle from the point set, this paper first obtains the convex hull of the point set. The convex hull of a finite point set S is the set of all convex combinations of its points. In a convex combination, each point x_i in S is assigned a weight or coefficient α_i in such a way that the coefficients are all non-negative and sum to one, and these weights are used to compute a weighted average of the points. For each choice of coefficients, the resulting convex combination is a point in the convex hull, and the whole convex hull can be formed by choosing coefficients in all possible ways. Expressing this as a single formula, the convex hull is the set:

$$Conv(S) = \left\{\sum_{i=1}^{|S|} \alpha_i x_i \,\middle|\, (\forall i: \alpha_i \geq 0 \wedge \sum_{i=1}^{|S|} \alpha_i = 1)\right\} \tag{2}$$

For instance, when S is a bounded subset of the plane, the convex hull may be visualized as the shape enclosed by a rubber band stretched around S. Convex hull of a point set is shown in figure 6.

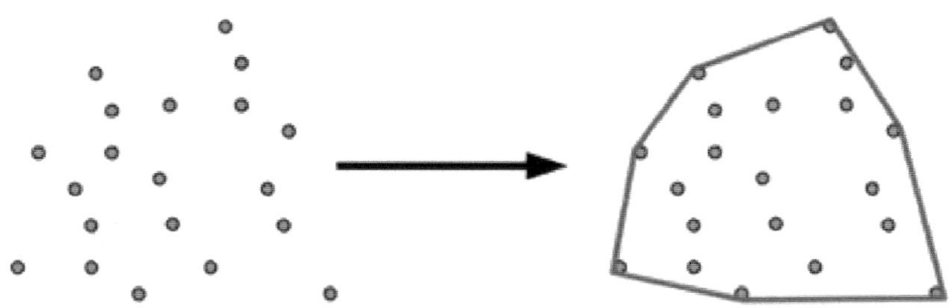

Figure 6. Convex hull of a points set

In this paper, the convex hull of the gear contour is the smallest convex polygon that can contain the contour of the gear, which is acquired by the GCH algorithm (Sklansky 1982). And the vertexes of the polygon are distributed on the top of each tooth. For the gear contour in figure 5(c), it’s convex hull is shown in figure 7.

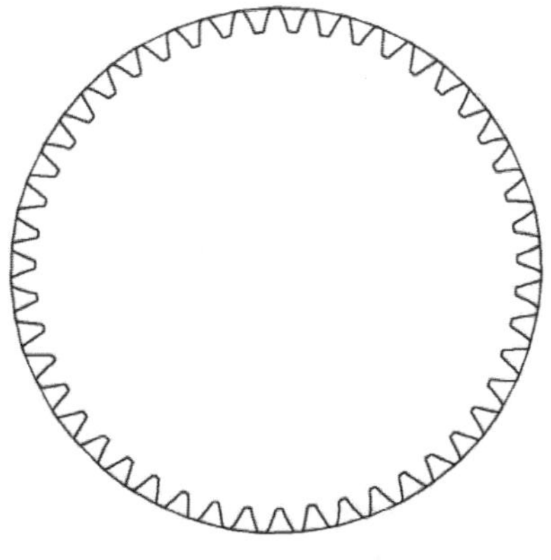

Figure 7. Convex hull of the gear contour

2.4. Acquisition of the convexity defects

Any deviation of the object from this hull can be considered as convexity defect. That means an area that do not belong to the object but located inside of its convex hull. Expressing convexity defects as a formula.

$$ConvexityDefects = Conv(Contour) - Contour \tag{3}$$

In the field of machine vision, convexity defects are often used for gesture recognition and limb motion recognition. For convenience, convexity defects can be represented by a quaternion which contains these values: [start, end, farthest, depth].

$$Defects = \sum \Delta(start, end, fasthest) \tag{4}$$

Start and end denote the start and end points of the defect area on the contour, corresponding to point A and point B in figure 8, respectively. Farthest represents the point furthest from the convex hull in the defect area, corresponding to point C in figure 8. Depth represents the shortest distance between the farthest point and the convex hull, which is shown in figure 8 by the line with arrow. For the sake of convenience, this paper uses the triangle consisting of point A, B, C to describe a defect.

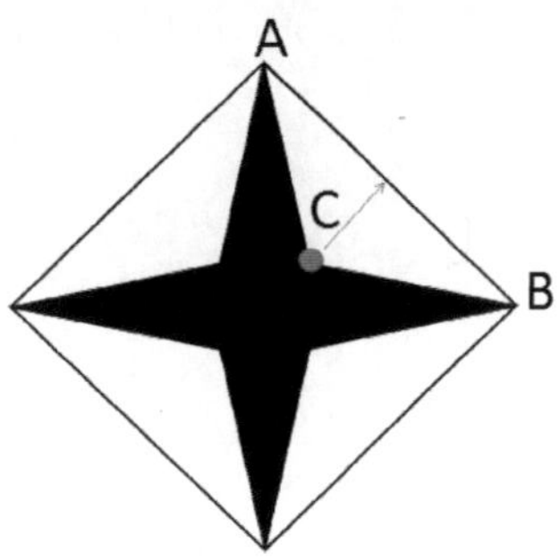

Figure 8. Convexity defects

In this paper, a defect between the contours of the gear and the convex hull is shown below in figure 9. The starting point A and the end point B of the defect are located on the top of the adjacent teeth. In other words, point A and B are located on the addendum circle. Obviously, the farthest point C is

IOP Conf. Series: Journal of Physics: Conf. Series **1183** (2019) 012010 doi:10.1088/1742-6596/1183/1/012010

located on the dedendum circle.

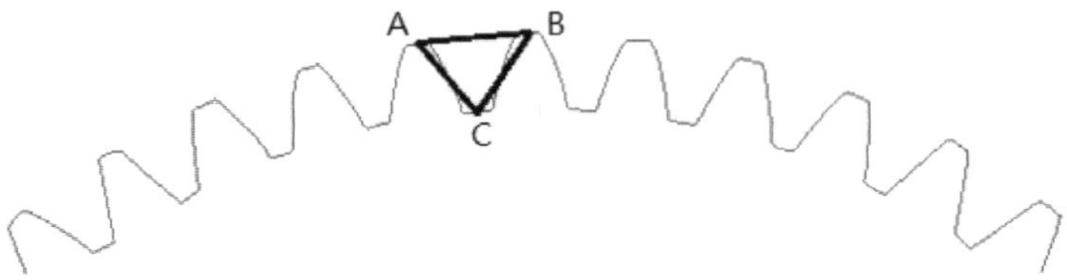

Figure 9. Triangle ABC presents a defect between the contours and the convex hull.

3. Fast calculation of gear parameters

The use of contour extraction of the new method proposed in this paper, instead of the edge extraction process in traditional methods, omits the need to extract point sets from edge images. After obtaining the convex hull and the convexity defect of the gear contour, extract the limited number of key points of the gear profile. The process greatly reduce the amount of calculation, one can quickly calculate the gear parameters.

3.1. The number of the gear teeth

According to the nature of the gear, the number of gear teeth is equal to the number of defects, also equal to one-third of the number of key points.

$$Num(teeth) = Num(defects) \tag{5}$$

3.2. Radius of the dedendum circle and addendum circle

After obtaining the defects between the contour and the convex hull, the starting point and the end point of each defect triangle are on the crest circle, and the farthest point is on the root circle. The minimum circumscribed circles is fitted according to the starting point (or end point) point set and the farthest point set, respectively.

The radius of the minimum circumscribed circle of the starting point (or end point) set is equal to the tooth top circle radius, and the radius of the minimum circumscribed circle of the farthest point set is equal to the radius of the tooth root circle.

The dedendum circle and addendum circle obtained by fitting the key points of the gear contour in figure 5(c) is shown in figure 10.

The radius is determined by the coordinates in the image, measured in pixels. Once the camera is calibrated, the number of pixels can be converted to the actual unit of measure, thus the radius can be expressed in actual dimensions.

Figure 10. The dedendum circle and addendum circle

IOP Conf. Series: Journal of Physics: Conf. Series **1183** (2019) 012010 doi:10.1088/1742-6596/1183/1/012010

4. Experiment

The experiment consists of two parts, hardware and software. Hardware includes desktop computer with Windows operating system, two megapixel DaHeng Mercury GigE series industrial camera (MER-200-14GM). As for software part, the algorithm is implemented in C++ based on Visual Studio and data processing is done by Python. This paper compares the new methods proposed in this paper and the classical methods proposed in paper [9].

The gear diameter measured by the vernier caliper is treated as the standard parameter, which is compared with the results obtained by the new method and the classical method, respectively. The results are shown in table 1.

Table 1. Results obtained by the new method and the classical method

	Radius of addendum circle(mm)	Radius of dedendum circle(mm)	Gear teeth number
standard value	17.86	15.78	34
new method	17.8879	15.8405	34
classical method	17.8879	15.3017	34

Both methods are performed 50 cycles, then record the run time. Finally draw a line chart, shown below in figure 11.

As can be seen from table 1, for the top circle, the measurement errors obtained by the two methods both are 1.56 ‰, and for the root circle, the measurement error obtained by our method is 3.82‰ and 3.13% is achieved by the classical one.

As for the time consumed for 50 running cycles, the new method is 1.97 times less than the classical one, meaning nearly twice improvement of the performance.

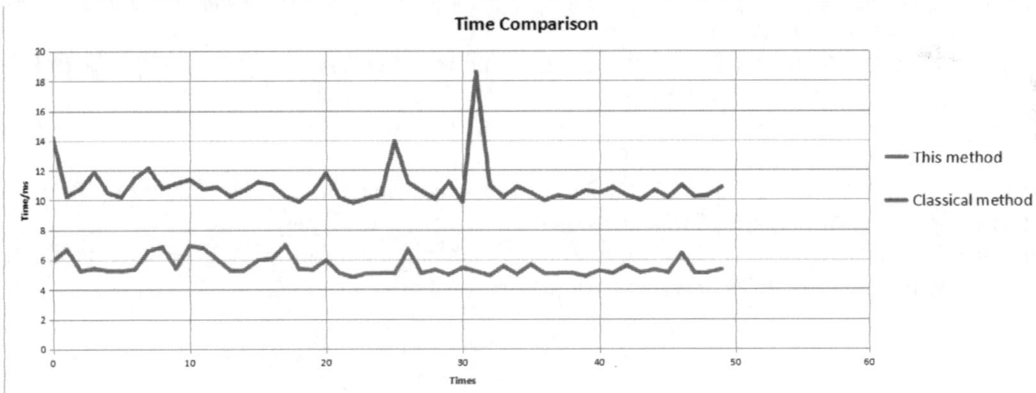

Figure 11. Time contrast of performing 50 times.

5. Conclusion

This paper presents a method for rapid measurement of gear parameters based on contour convex hull and convexity defects. Firstly, a series of pre-processing, including gray-scale transformation, threshold segmentation and contour extraction, is performed on the gear image. And then obtains the finite key points on the gear profile by finding the convexity defects of the gear contour. After that, gear parameters like gear teeth number, radius of the dedendum circle and addendum circle can be calculated quickly and easily. In the process of image preprocessing, the new method uses the contour extraction instead of edge detection to get the point set of gear contour. So the operation of extracting a set of points from the edge image is omitted. Parameter calculation based on key points makes it unnecessary for the new method to have all the points of the gear profile involved in the calculation, which greatly reducing the amount of calculation. The comparative experiments show that the new method speeds up nearly two times in the case of maintaining the same accuracy, compared with the classic method.

References

[1] Chen M and Li Y 2010 The study of gear Parameter Measurement method based on machine vision Proc. 2010 Int. Conf. on Intelligent Control and Information Processing (Dalian, China, 13–15 August 2010) (New York: IEEE) pp 248–251.
[2] Ge D, Yao X, Lian Z, Wang H and Tao T 2017 Applications of computer vision in measuring total cumulative pitch deviation of a gear *Tehnički vjesnik* **24** 71–8.
[3] Klancnik S, Ficko M, Balic J and Pahole I 2015 Computer vision-based approach to end mill tool monitoring *Int. J. Simul. Model.* **14** 571–83.
[4] Kovalenko M, Antoshchuk S and Sieck J 2015 Real-time hand tracking and gesture recognition using semantic-probabilistic network. *Proc. 2014 UKSim-AMSS 16th Int. Conf. on Computer Modelling and Simulation (UKSim) (Cambridge, UK, 26–28 March 2014)* (New York: IEEE) pp 269–274.
[5] Li P and Liu Q H 2009 Dimension measurement of straight-cut gear based on machine vision *Comput. Meas. Control* **17** 646–8.
[6] Lin C, Zeng D, Zhao X L and Cao X J 2015 Numerical calculation of tooth profile of a non-circular curved face gear *J. Mech. Eng.* **61** 303-10.
[7] Ma S D and Zhang Z Y 1998 *Computer Vision Computational Theory and Algorithm Base* (Beijing: Scientific Press).
[8] Simunovic G, Svalina I, Simunovic K, Saric T, Havrlisan S and Vukelic D 2016 Surface roughness assessing based on digital image features *Adv. Prod. Eng. Manag.* **11** 93-104.
[9] Sklansky J 1982 Finding the convex hull of a simple polygon *Pattern Recognit. Lett.* **2** 79–83.
[10] Suzuki S, Be K 1985 Topological structural analysis of digitized binary images by border following *Comput. Gr. & Image Process.* **30** 32-46.
[11] Wang D Z and Mottershead J E 2016 Measurement precision and spatial resolution with kriging digital image correlation *J. Mech. Eng.* **62** 419–29.
[12] Wang W C 2011 Design of measurement system of gear parameter based on machine version *J. Mech. Transm.* **35** 41–3.
[13] Yin X C, Zhao F C, Yin Y J and Li L L 2013 Research of machine vision technology in the wear test of gear *Adv. Mater. Res.* **718-720** 908–11.
[14] Zhang S J, Gou Z K, Li, Li Q L, Li Z F and Jin J 2004 Digital image processing technology for spur gear measurement *Optics Precis. Eng.* **12** 619–25.

IOP Conf. Series: Journal of Physics: Conf. Series **1183** (2019) 012011 doi:10.1088/1742-6596/1183/1/012011

Transparent object detection and location based on RGB-D camera

Chen Guo-Hua, Wang Jun-Yi and Zhang Ai-Jun
Beijing University of Chemical Technology

E-mail: chengh@mail.buct.edu.cn

Abstract. In order to improve the accuracy and efficiency of robot grasping, we propose a new method for transparent object detection and location that utilize depth image, RGB image and IR image. In detection process, an active depth sensor (RealSense) is firstly employed to retrieve the transparent candidates from the depth image and the corresponding candidates in the RGB image and IR image are then extracted separately. A transparent candidate classification algorithm is subsequently presented that uses SIFT features to recognize the transparent ones from the candidates. In location process, we obtain a new group of RGB images and IR images by adjusting camera orientation to make its optical axis perpendicular to the normal direction of the plane on which the object is placed. The object contours in RGB image and IR image are then extracted, respectively. The three-dimensional object is finally reconstructed by means of stereo matching of the two contours, and the current pose information of the object is calculated in the end. In order to verify the feasibility of the method, we built a hand-eye test system with a movable industrial robot to detect and capture transparent objects at different locations. The final test results demonstrate that the method is more general and effective than the traditional one.

Keywords: Transparent Object Detection, location, RGB-D Camera, Depth Image, Robotic Grasping.

1. Introduction

Transparent objects are very common in our daily life, such as glasses and bottles in the home, beakers and test tubes in the laboratory. With the development of robotics, robots have been widely used in many fields, such as service robots in the domestic environment and industrial robots in the laboratory environment. These robots inevitably involve the identification and grasping of transparent objects. Particularly in unmanned laboratories, transparent object identification and grasping is especially a key issue.

Object recognition has been a hot research topic in computer vision, but transparent object recognition is a difficult problem to solve, because transparent materials are difficult to detect due to the appearances of transparent objects that can change over different backgrounds. We use the advantage of the Kinect sensor, which provides scene depth information using a projected pseudo-random IR pattern, but it cannot form effective depth data on the transparent object surface.

In order to solve this problem, we present the detection and pose recognition of transparent objects based on RealSense's RGB-D information and movable robotic arm and we propose several methods in this paper. Firstly, describe the previous work has been done. Secondly, we propose algorithm to segmentation and recognition. Next, we propose a new method to reconstruction the objects. Then, we use the model to pose estimation of rigid transparent objects. Finally, we evaluate the proposed approach by using this method in five objects and investigate performance in various scenes containing transparent objects.

2. Related work

There are three types of recognition method of transparent objects: 2D Computer Vision, 3D Computer Vision and others.

2D Computer Vision. The appearance of a transparent object strongly depends on lighting and its background. Transparent objects usually don't have their own texture features, their edges are typically weak and intensity gradient features are heavily influenced by seeing through background clutter as well as secularity edges induced by lighting. So that classical computer vision algorithms for recognition and pose estimation are difficult to apply to transparent objects. To solve these problems, an Additive Latent Feature Model is used for Transparent Object Recognition as [1] showed, they used a LDA-SIFT model, [2] proposed a new feature called the light field distortion (LFD) feature. Based on [2], [3] showed a LFD feature and use it to segment the transparent objects from RGB image, and it got a better result. [4] showed a video of a transparent object shot under varying illumination to estimate the normal map, [5] combined the systematic distortions in background texture occurring at the boundaries of transparent objects with the strong lightened typical of glass surfaces to train a hierarchy of classifiers, identify glass edges, and to find consistent support regions for these edges. [6] employed a probabilistic formulation for segmentations glass regions, similar to the [5], [7] proposed to use binary criteria to guide the segmentation process, [8] reconstructed a transparent surface modelling from a pair of polarization images.

3D Computer Vision. 3D point clouds are successfully used for object recognition and pose estimation. However, modern sensors (Kinect, ToF cameras, stereo cameras, laser scanners) can't estimate depth reliably and produce point clouds for transparent and specular objects so these algorithms cannot be applied. Cross-modal stereo can be used to get depth estimation on transparent objects with Kinect as in [14], but its quality is inferior. Acquisition of 3D data and reconstruction of transparent objects is a challenging and unsolved problem. [14] showed the method of using missing depth information to segment the object and pose estimate which first proposed by [19], [16] used Geometric Hashing to solve the clutter. [12] reconstructed objects by moving around a transparent object in the scene, [13] combined multiple sensor modalities, [15] showed a method of using seashell sensors to add missing point cloud data. [19] improved the Cross-modal stereo by using fully-connected CRF.

There also have some other methods by using non-image sensors, such as using colour image and laser reflectance image that extracting object via reflectance laser intensity, or using a stereo camera and three-dimensional laser radar system to simultaneously illuminate transparent objects by a plurality of light sources and identify transparent objects by the high spot.

But, the quality of transparent object recognition by using traditional RGB information is highly dependence on the transparency of the object and the ambient light, and these methods also have certain limits of shooting angle of an object and cannot obtain the depth information form the object. And the method that uses non-image sensor is very complex that cannot suitable for the needs of daily life. The traditional use of stereo visual recognition needs to advance the identification of objects such as modelling operations. So the traditional identification process is very complex.

3. Recognition of Transparent Objects

RealSense uses the laser speckle pattern technique to estimate the depth, it can project infrared light onto the surface of the object. However, it cannot form an effective pattern on the surface of transparent objects and also cannot get the effective point cloud information. Based on this condition, we can use the area where has no depth information as the region of interesting in depth image, and segment this region in the RGB image. Then we use the SIFT feature to detect the change of the feature points of the segmented region. According to the movement of feature points to determine whether the object is transparent.

IOP Conf. Series: Journal of Physics: Conf. Series **1183** (2019) 012011 doi:10.1088/1742-6596/1183/1/012011

3.1. Transparent Candidate Retrieval

The method we used is based on [11], but we improved the method to exclude non-transparent objects which can generate invalid depth areas. RealSense not only cannot estimate the depth of transparent objects, and it also cannot estimate contours of objects. Sometimes it can receive the depth information of the background behind transparent objects as figure 1, but we can take advantage of this fact: regions where RealSense cannot form effective depth information are likely to belong to transparent objects. So this cue can be used to segment transparent objects on an image. We can use this cue to generate the mask of transparent objects. And we use GrabCut segmentation algorithm on RGB image, we can get the contours and area of the object as figure 2.

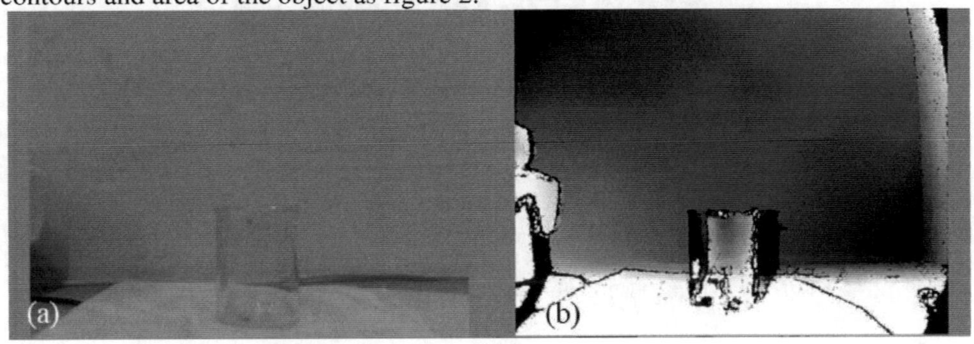

Figure 1. (a) RGB image (b) Depth image

Before we get the interesting region of the depth image, firstly we use the formulas as (1) and (2) to the depth image to remove the noise caused by the infrared beating of the device itself.

$$\mathrm{w} = w_g \times w_c \tag{1}$$

$$w_c = exp\left(-\frac{\left(G(i,j)-G(x,y)\right)^2}{2\sigma_c^2}\right) \tag{2}$$

Where (i, j) is the standard deviation of the Gaussian function, and G (i, j) is the standard value when the depth image is transformed into a grayscale image at the pixel (i, j)) at the grey value.

After filtering, morphological operations (closing and opening) are applied to find and eliminate small regions in the depth image, the result is a mask of transparent objects (figure 2(a)-2(b)). Using the mask information, the Grab Cut segmentation algorithm is used to segment the area of the transparent object on the corresponding RGB image and IR image. We can use the mask information as the initial conditions and constraints of the algorithm when we use Grab Cut, we can get the final segmentation results that are shown in figure 3.

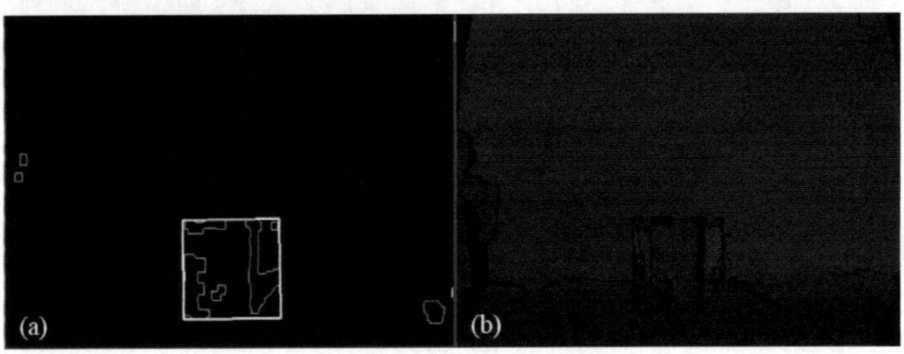

IOP Conf. Series: Journal of Physics: Conf. Series **1183** (2019) 012011 doi:10.1088/1742-6596/1183/1/012011

Figure 2. (a) Mask of depth image (b) Depth image

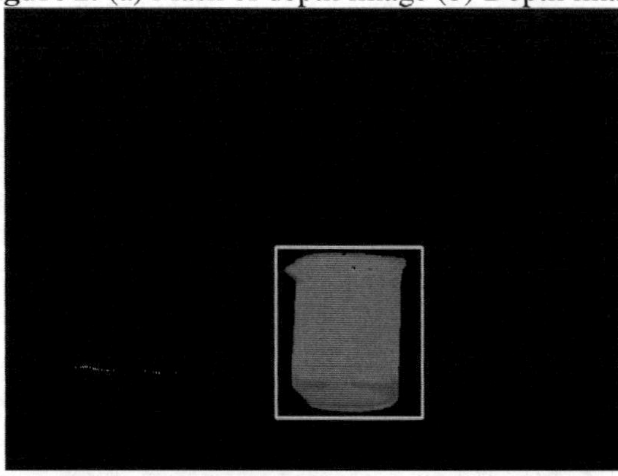

Figure 3. Results of glass segmentation

3.2. Transparent Candidate Classification

Transparent objects are made of refractive materials; such as glass or plastics. It can distort rays emanating from the scene background (figure 4). We use the phenomenon of background distortion caused by refraction to identify transparent objects.

Figure 4. Background distortion from changing viewpoints

We use the SIFT feature to classification the transparent candidate. As described in figure 5, due to refraction of the transparent object, background texture changes irregularly. Thus, the SIFT features obtain from the transparent object are more irregular than features from the opaque object, and these features deviate from the hyperplane due to the Lambertian reflection in the phase space. We could use optical flow method to detect the feature points. When we move the camera linearly, if the feature points move irregularly as figure 5, the area which the key points in is the transparent area.

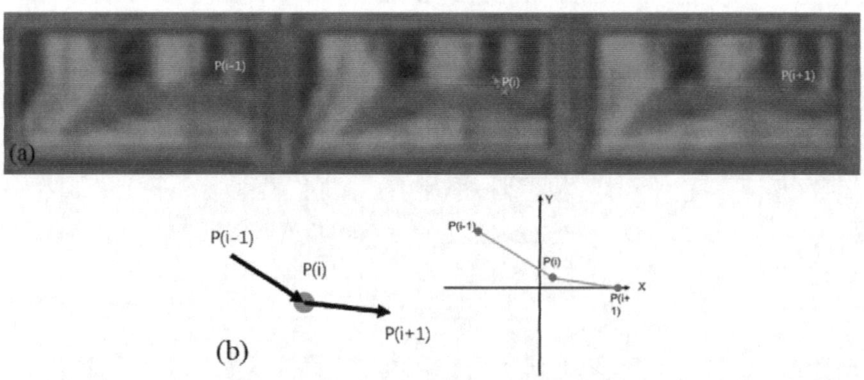

Figure 5. (a) The feature points extracted from picture (b) The feature points move irregularly

We take an arbitrary feature point in the image from one of the viewpoints as $p_s(0)$, and the next few frames of this feature point as $p_s(i)$. And we use the following formulae to express the change of this viewpoint,

$$\Delta p_s(i) = p_s(i+1) - p_s(i) \tag{3}$$

$$\sigma_s{}^2 = \sum\left(\Delta p_s - \widehat{\Delta p_s}\right)^2 / M \tag{4}$$

$$K = \sum\left(\sigma_s{}^2 - \widehat{\sigma^2}\right)^2 / N \tag{5}$$

As shown in these formulae, K is the threshold. We can adjust the value of K to determine whether the region of interest is transparent, M is the number of frames that can continuously stabilize the feature points, N is the number of feature points that can continuously and steadily obtain.

4. Reconstruction

In the modelling phase, we first adjust the camera's angle, the camera axis and the object plane perpendicular to the use of cross-mode stereo matching, in multiple locations to obtain RGB images and IR images, extract the image of the object contour information on the object reconstruction. Finally, the reconstructed model is used to estimate the pose of the object.

4.1. Contour extraction

Firstly, we adjust the camera's axis position perpendicular to the normal direction of the plane where the object is located, and then we use the method mentioned earlier to segment the object to divide the contours from the RGB image and the IR image. As shown in figure 6, the object contour edge is very rough. In order to ensure that the reconstruction results are accurate, we smooth the contour information. Since the Gaussian distribution has good characteristics, and the Gaussian convolution can suppress the size of the noise by controlling the size of σ. so we use the Gaussian filter for filtering contour.

$$G(t,\sigma) = k\sigma^2 exp\left(-\frac{t^2}{2\sigma^2}\right) \tag{6}$$

In this formula, K is a constant, we can adjust the value of k to change the filter coefficients. The filter window typically is 6σ~8σ.

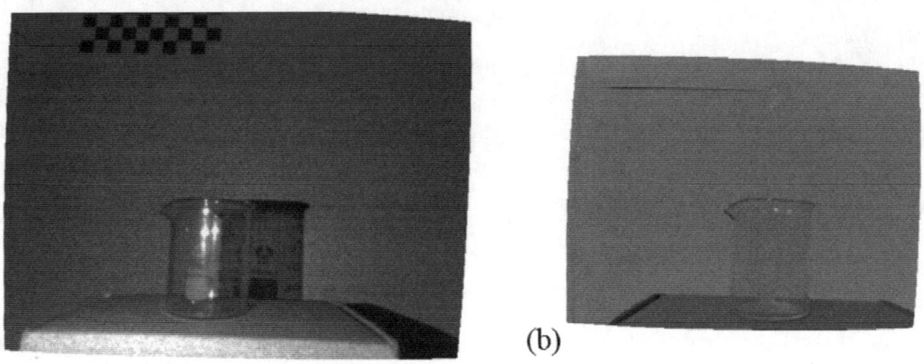

(a) (b)

IOP Conf. Series: Journal of Physics: Conf. Series **1183** (2019) 012011 doi:10.1088/1742-6596/1183/1/012011

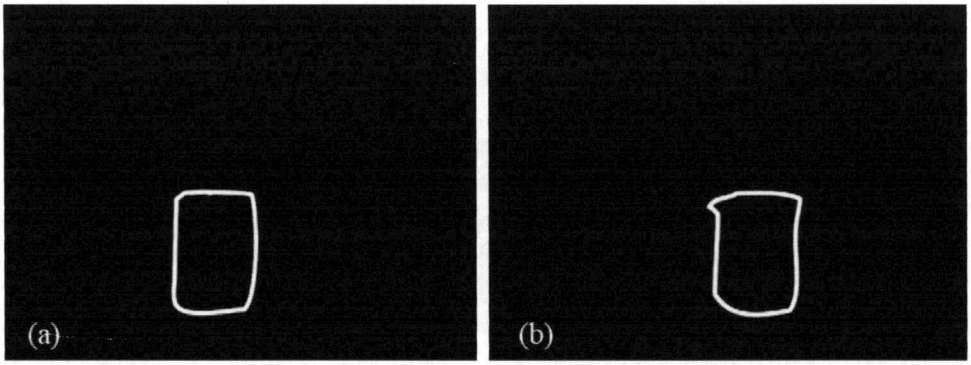

Figure 6. (a) IR image (b) RGB image (c) Contours of IR image (d) Contours of RGB image.

After we get the contours obtained from the two views, we need to fit the contours. Since the contours in the IR image are more flawed than the RGB images, so we can merge these two contours together and get the maximum of the contours. Firstly, we obtain the central moment of the two contours, then merge the central moments, finally we extract the outer parts of the two contours. As shown in figure 7.

Figure 7. Merge of contours

4.2. Reconstruction

While moving the device, we extract contours from IR images and RGB images to use stereo matching to obtain the approximate mean plane information and the axis position information of the object as a three-dimensional reconstruction of the object geometric information. And we can get the object contour normalized by rotating, and get the three-dimensional information of the object. As shown in figure 8.

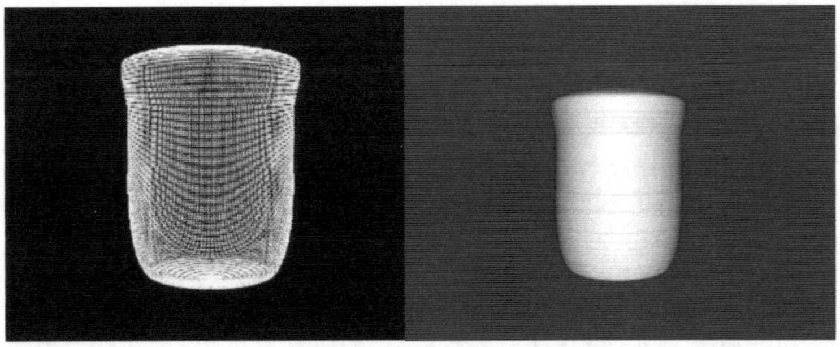

Figure 8. Point cloud data and surface reconstruction

4.3. Objects Capture

After these previous steps, we already get the object of the three-dimensional model information and the current location of the object. Because we need to capture transparent objects in the Lab finally, so we need a method to control the robot arm to capture the object accurately.

In order to make the robot grasp the object accurately, we refer to many papers of robot grasping such assume of the object's current position information, via the outline of the object to find the appropriate two points to capture and carry out effective capture behaviour. The complete process is shown in figure 9.

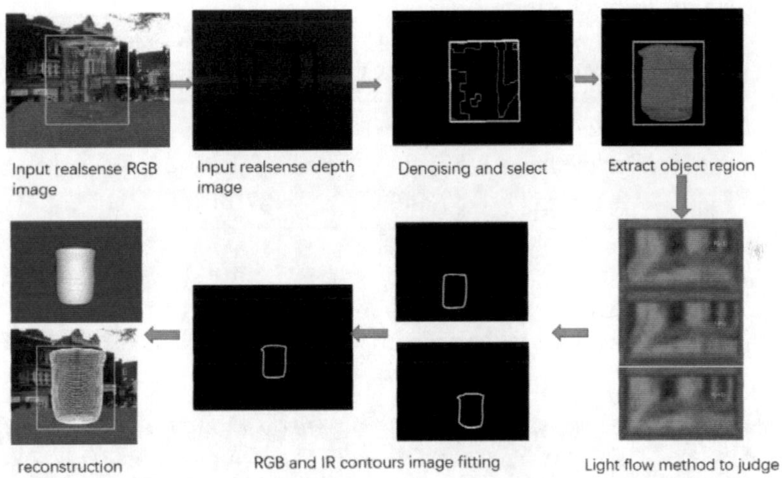

Figure 9. Process of reconstruction and pose estimation

5. Experiment

Here we use RealSense sensor to get the image, this sensor has a colour sensor (1080P@30FPS) and a depth sensor (640X480@60FPS). We fix the camera at the robotic arm to test this method.

5.1. Assumption

We evaluated our proposed method by capture the transparent objects using RV-4F robot in a laboratory setting and real environments under following assumptions;

1 the object is not far from the robot in order to do this task without moving the base.

2 that there is no mutual occlusion between transparent objects

3 the poses of transparent objects are fixed after the experiment begins.

4 the background of the object is grained.

IOP Conf. Series: Journal of Physics: Conf. Series **1183** (2019) 012011 doi:10.1088/1742-6596/1183/1/012011

We performed some experiments in a laboratory and real setting to evaluate robustness and limitations of our proposed method

5.2. Recognition of transparent objects

We test the performance of transparent object recognition and compare the results with the method proposed in [6]. In this test, we use accuracy and miss to evaluate the performance of transparent object recognition. Accuracy is the percentage of the number of correctly recognized transparent object over the number of all tested transparent objects. Miss is the ratio of object that is opaque, doesn't have depth information and misrecognized as an opaque one.

Miss refers to the proportion of the opaque objects that are mistaken as transparent ones in all the opaque objects. The main reason why opaque objects may be considered as transparent ones is that some opaque objects are specular objects. The ambient light, therefore, may be strongly reflected by the smooth surface during the detection process. The reflected environmental light information is regarded as the background information of the occlusion area of the object, and thus the object is consequently judged to be a transparent object.

We randomly put five transparent objects (glass goblet, beaker, test tube, cups and graduate) and other non-transparent objects that will cause unknown depth value in the scene to test the miss. In the test, we will put transparent objects and non-transparent objects in the same scene. The results of our tests are shown in table 1. As can be seen in table 1, both methods have high recognition rates of the transparent object area, but our method is better when identifying non-transparent objects. Because our method can recognize more accurate than traditional one in non-transparent objects recognition which have unknown value in the depth image.

Table 1. Accuracy and miss of transparent object recognition

Method	Accuracy	Miss
Our method	87.64%	37.93%
Method in [11]	87.64%	87.24%

5.3. Object Reconstruction

Finally, in figure 10 we show the result of our approach to reconstruct 3D models. In order to test the accuracy of our pose detection, we use the robot to capture the objects. We successfully capture 70 out of 85.

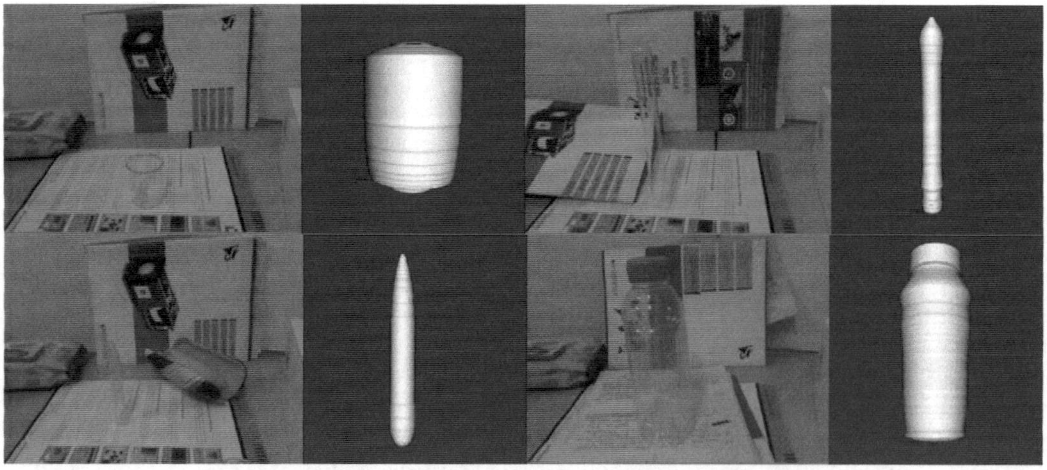

Figure 10. Left: Real scene. Right: Reconstructed model

5.4. Distance Analysis

Since the depth sensor has a range of work, so we move the camera in a difference of ±15cm from the reference position 60cm. Figure 11 shows that the capture ratios are decreased when the displacement further from the reference position, because the accuracy of depth information of objects that sensors detect varies due to the distance.

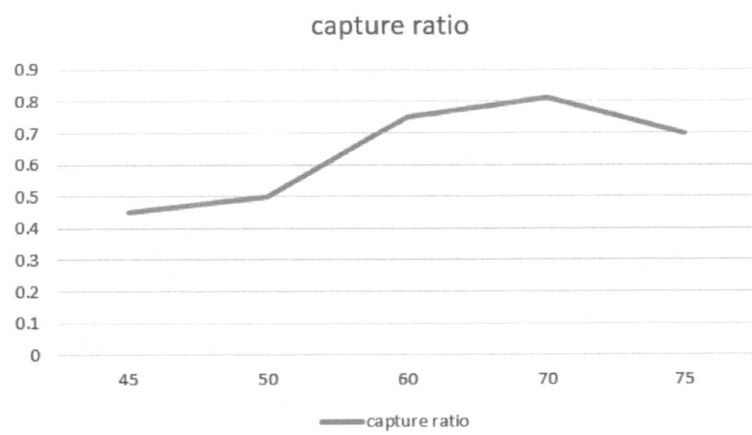

Figure 11. The relationship between Distance and grab success rate.

6. Conclusion

Using this method, the rotating transparent objects can be effectively detected and captured, and the 3D models of the transparent objects are built meanwhile. These 3D models will be saved into a database and used as source materials in the future recognition process, for example, they can be used as templates to deal with the occlusion situation in the geometric Hexa method as mentioned before [11]. Additionally, this method can be applied for autonomous mobile robots to avoid transparent obstacles. The shapes of identified transparent objects of this method are limited, for it can only deal with the rotating objects, hence in next step we will focus on expanding the identifying scope of shapes.

References

[1] Fritz M, Bradski G, Karayev S, Darrell T and Black M J 2009 An additive latent feature model for transparent object recognition *Proc. DBLP Advances in Neural Information Processing Systems (Vancouver, Canada, 7–10 December 2009) (Red Hook, NY: Curran)* pp 558–566.

[2] Maeno K, Nagahara H, Shimada A and Taniguchi R I 2013 Light field distortion feature for transparent object recognition. *Proc. IEEE Int. Conf. On Computer Vision and Pattern Recognition (Portland, OR, 23–28 June 2013) (New York: IEEE)* pp 2786–2793.

[3] Xu Y, Nagahara H, Shimada A and Taniguchi R I 2015 TransCut: Transparent object segmentation from a light-field image. *Proc. IEEE Int. Conf. On* Computer Vision *(Santiago, Chile, 7–13 December 2015) (New York: IEEE)* pp 3442–50.

[4] Yeung S K, Wu T P, Tang C K, Chan T F and Osher S J 2015 Normal estimation of a transparent object using a video *IEEE Trans. Pattern Anal. Mach. Intell.* **37** 890–7.

[5] McHenry K, Ponce J and Forsyth D 2005 Finding glass. Proc. IEEE Int. Conf. On Computer Vision and Pattern Recognition (San Diego, California, 20–26 June 2005) (vol 2) (New York: IEEE) pp 973–9.

[6] Torres-Gómez A and Mayol-Cuevas W 2014 Recognition and reconstruction of transparent objects for augmented reality. *Proc. IEEE Int. Sympos. On Mixed and Augmented Reality (ISMAR) (Munich, Germany, 10–12 September 2014)* (New York: IEEE) pp 129–34.

[7] McHenry K and Ponce J 2006 A geodesic active contour framework for finding glass. Proc. IEEE Int. Conf. On Computer Vision and Pattern Recognition (New York, NY, 17–22 June 2006) vol1

(New York: IEEE) pp 1038–44.
[8] Miyazaki D, Kagesawa M and Ikeuchi K 2004 Transparent surface modeling from a pair of polarization images *IEEE Trans. Pattern Anal. Mach. Intell.* **26** 73–82.
[9] Klank U, Carton D and Beetz M 2011 Transparent object detection and reconstruction on a mobile platform. Proc. IEEE Int. Conf. On Robotics and Automation (ICRA) (Shanghai, China, 9–13 May 2011) (New York: IEEE) pp 5971–78.
[10] Eren G, Aubreton O, Meriaudeau F, Secades L S, Fofi D, Naskali A T, Truchetet F and Ercil A 2009 Scanning from heating: 3D shape estimation of transparent objects from local surface heating *Opt. Express* **17** 11457–68.
[11] Lysenkov I, Eruhimov V and Bradski G 2013 Recognition and pose estimation of rigid transparent objects with a kinect sensor *Proc. Robot. Sci. Syst.* **8** 273–5.
[12] Alt N, Rives P and Steinbach E 2013 Reconstruction of transparent objects in unstructured scenes with a depth camera. Proc. IEEE Int. Conf. On Image Processing (ICIP) (Melbourne, Australia, 15–18 September 2013) (New York: IEEE) pp 4131–35.
[13] Hagg A, Hegger F and Plöger P 2016 On recognizing transparent objects in domestic environments using fusion of multiple sensor modalities *Preprint* arXiv/1606.01001.
[14] Chiu, W W C, Blanke U and Fritz M 2011 Improving the kinect by cross-modal stereo. *Proc. Int. Conf. On British Machine Vision (Dundee, Scotland, 29 August–2 September 2011)* (Manchester, UK: BMVC Press) p 3.
[15] Huang K, Jiang L T, Smith J R and Chizeck H J 2015 Sensor-aided teleoperated grasping of transparent objects. *Proc. IEEE Int. Conf. On Robotics and Automation (ICRA) (Seattle, WA, 26–30 May 2015)* (New York: IEEE) pp 4953–9.
[16] Lysenkov I and Rabaud V 2013 Pose estimation of rigid transparent objects in transparent clutter. *Proc.IEEE Int. Conf. On Robotics and Automation (ICRA) (Karlsruhe, Germany, 6–10 May 2013)* (New York: IEEE) pp 162–9.
[17] Lei Z, Ohno K, Tsubota M, Takeuchi E and Tadokoro S 2011 Transparent object detection using colour image and laser reflectance image for mobile manipulator. *Proc. IEEE Int. Conf. On Robotics and Biomimetics (ROBIO) (Karon Beach, Thailand, 7–11 December 2011)* (New York: IEEE) pp 1–7.
[18] Saygili G, van der Maaten L and Hendriks E A 2014 Hybrid kinect depth map refinement for transparent objects. *Proc. IEEE Int. Conf. On Pattern Recognition (Stockholm, Sweden, 24–28 August 2014)* (New York: IEEE) pp 2751–6.
[19] Ulrich M, Wiedemann C and Steger C 2009 CAD-based recognition of 3D objects in monocular images. *Proc. IEEE Int. Conf. On Robotics and Automation (ICRA) (Kobe, Japan, 12–17 May 2009)* (New York: IEEE) pp 1191–8.
[20] Lin Y C, Wei S T and Fu L C 2014 Grasping unknown objects using depth gradient feature with eye-in-hand RGB-D sensor. *Proc. IEEE Int. Conf. On* Automation Science and Engineering (CASE) *(New Taipei, Taiwan, 18–22 August 2014)* (New York: IEEE) pp 1258–63.

IOP Conf. Series: Journal of Physics: Conf. Series **1183** (2019) 012012 doi:10.1088/1742-6596/1183/1/012012

Evaluation method of running performance for five-axis machining center based on the "S" specimen

H J Wang[1,2], F X Han[1], Y H Gu[2], B G Rosén[2,3] and A N Zou[1,2]
[1]School of Mechanical and Electrical Engineering, Beijing Information Science and Technology University, China.
[2]Key Laboratory of Modern Measurement and Control Technology, Ministry of Education, Beijing Information Science and Technology University, China.
[3]The Rydberg laboratory, Halmstad University, Halmstad, Sweden.

E-mail: wanghongjun@bistu.edu.cn

Abstract. With the development of industries and the advanced manufacturing technology, the performance of the machine tools plays an important role for the product quality. Machining performance of five-axis machining centers, MC, in the manufacturing industry has been a hot research area, where evaluation methods based on cutting of a test piece are common, but it have proved hard to determine the relationship between test piece deviations and machine tool sensor data during manufacturing. An "S" test specimen has the characteristics of open and close angle, variable curvature, thin wall, and low stiffness. The quality of a machined "S" specimen is used to test the characteristics of five-axis MCs. Inaccurate manufacturing results in unqualified parts and the loss of business. Normally geometrical errors are hard to adjust and compensate for in the process. In order to evaluate, and map the relationship between the characteristics of geometrical work-piece deviations from a MC to in-line sensor data, a monitoring system was established in this study. The LABVIEW monitoring evaluation system is developed based on "running performance tracing". By using a coordinate measuring machine, CMM, and a Scanning Electron Microscope to measure the actual geometry and surface finish of the machined S part and compare it to the ideal geometry, a database and a running performance evaluation model for 5-axis MC based on the "S" specimen was developed. Finally, the effectiveness of the method of evaluation is verified by cutting experiments using a MAZAK five-axis MC. The Achievements reported here are helpful for solving the lack of systems to evaluate the running performance of five-axis CNC machine tools for high-end manufacturing industry in China, which has important application prospect and high economic value.

Keywords: Five-axis MC, "S" specimen, evaluation model, running performance, surface finish.

1. Introduction

With complex structure and high precision, five-axis MCs are used for machining complex parts. The running performance of a five-axis machining center (MC) plays an important role on the products' quality and productivity. The errors of a part are resulted by the comprehensive errors of multi- axis. It is difficult to evaluate the performance of a five-axis MC [1]. The evaluation of machine tools requires a great deal of time and sometimes does not reflect the accuracy of actual cutting. The standard parts for indirect measurement include NAS tester [2] and "S" Specimen. The "S" Specimen has the

IOP Conf. Series: Journal of Physics: Conf. Series **1183** (2019) 012012 doi:10.1088/1742-6596/1183/1/012012

characteristics of open and close angle, variable curvature, thin wall, low stiffness, which reflects not only the static errors but also the servo system tracking error, tool positioning error and profile error etc. The geometrical, kinematic characteristics of “S” Specimen have been investigated [3]. The map of stiffness on a surface has been proposed [4]. Zhong Jiang et al. [ref x] introduced an approach based on modeling and simulation of surface morphology [5].

The proposed positioning algorithm is applied to the S-shaped test [6]. The variation of stiffness lead to cutting deformation during an S tester, the vibration and the loss of accuracy have been analyzed [7]. The evaluating accuracy of a five-axis machine tool by measuring the finished tester has been widely studied [8-12]. However, the questions of how to evaluate the effect of processing performance in high speed operation and how to evaluate the performance of machine tools online in real time still remains to be answered.

In this study, a multi-sensor acquisition system for “S” Specimen processing is developed based on the “S” Specimen.

In addition, a mapping model of running performance for Multi-axis CNC Machine Tools based on “S” Specimen is proposed. “S” Specimens are divided into four typical regions, and the corresponding mapping of quality errors, -the surface finish and the dynamic signal characteristics of “S” Specimens is established. The performance evaluation method of the five axis machining centers applied to a MAZAK machining center. The shape and position errors of “S” Specimens were measured by a coordinate measuring machine, and the surface quality of the testers was analyzed by scanning electron microscopy. The validity of the performance evaluation method was proved to be valuable for evaluation of the performance of five-axis MC in real.

2. Geometric characteristics of “S” specimen analysis

“S” specimen in the current test is made from aviation-aluminum alloy (type of alloy). The projection of the upper and lower edge curves of the “S” Specimen on the base is shown in figure 1. S1 is the projection of the top edge on the base surface; S2 is the projection of the bottom edge line on the base plane; C is the intersection point of S1 and S2. Open angle and close angle of the “S” Specimen is shown in figure 2.The outer edge of the “S” Specimen greater than 90° is called “open angle”, less than 90° are called “close angle”. The C point is the turning point between open and close angle.

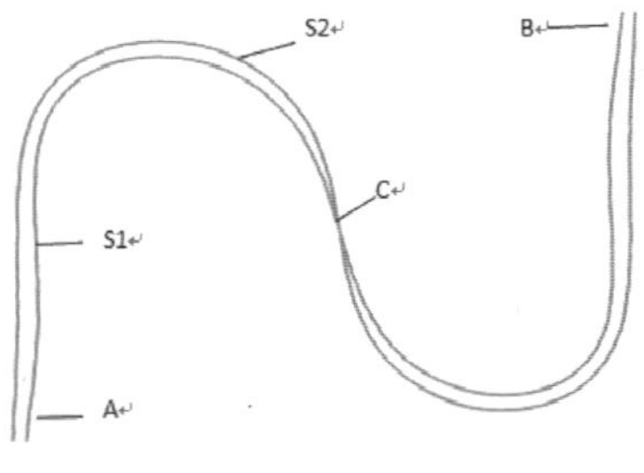

Figure 1. The projection of the upper and lower edge curves

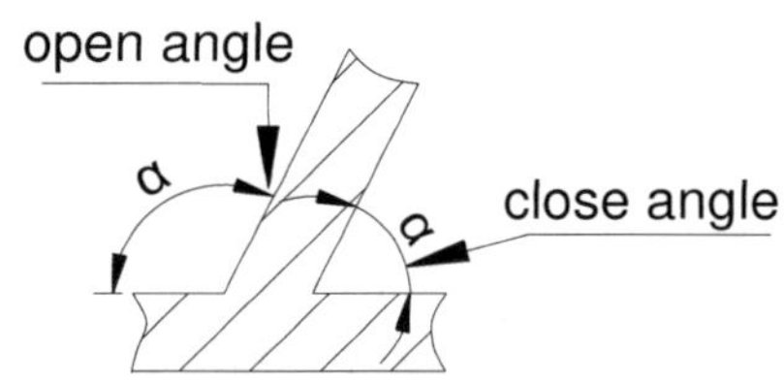

Figure 2. Machining angle defined

The curvature of “S” Specimen is shown in figure 3. The A-B curvature is almost linear, then sharply increasing to a bending curvature. After a maximum value, and constant reduction, through the intersection C, the curvature flips make a change direction. The continuity of the surface (shown in figure 4) is mainly G1 and G2 continuous. Because of “S” Specimen complexity structure, open and close angle, non-uniform changes during the machining process, the tool’s spatial violent posture changes, and the cutter axes vector always remains with the S side of the tangent. This enables the machine tool high multi-axis linkage accuracy and performance.

IOP Conf. Series: Journal of Physics: Conf. Series **1183** (2019) 012012 doi:10.1088/1742-6596/1183/1/012012

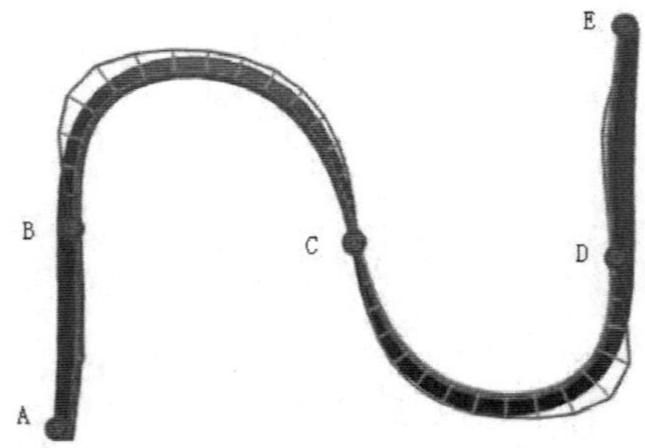

Figure 3. Curvature of the upper boundary

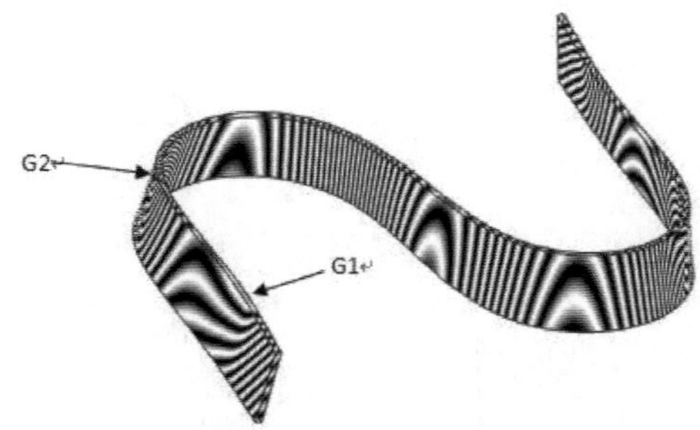

Figure 4. Continuity of surfaces

3. Evaluation model based on "S" Specimen precision and surface morphology

3.1. Influence of static performance of machine tool on accuracy of "S" Specimen

The geometric errors of MC include positioning error, straightness error, angular displacement error and squareness error. These errors affect the geometric accuracy of machine tool [13]. These errors will result in the oversized "S" Specimen base and S edge strip, shape and position errors, and sudden surface morphology changes.

3.1.1. Analysis of static accuracy of linear axis

The Kinematic error parameters of three linear axes X, Y and Z are shown in table 1. Taking the X-axis for example, 6 errors generated by the translation along the X axis include 3 linear errors and 3 rotational errors as shown in figure 5. That is, the location error $\delta_x(X)$ along the X-axis, the straightness error $\delta_y(X)$ along the Y-axis, and the straightness error $\delta_z(X)$ in the vertical direction along the Z-axis. In addition, see figure 6, there are three rotation errors ε_{xx}, ε_{yx}, ε_{zx}, and three squareness errors between the X-, Y- and Z-axis.

During machining the base of the S test piece, the rotating axis of the MC are fixed to A=0, and C=0. The processing methods of hole 1 and hole 2 are fine boring. Hole 1 is the reference hole, error δ_{xx}, δ_{yx} is determined by measuring the position of the hole 2 relative to the hole 1 as shown in figure 7. By measuring the verticality of the axis of the hole 1 and 2 axis on the upper surface A, the existence of squareness errors S_{yz}, S_{zx} is evaluated (see figure 8).

Table 1. Kinematic errors of the linear-axes

Description	Location errors	Straightness error	Angular displacement error
X-axis motion	$\delta_x(X)$	$\delta_y(X)$ $\delta_z(X)$	$\varepsilon_x(X)$ $\varepsilon_y(X)$ $\varepsilon_z(X)$
Y-axis motion	$\delta_y(Y)$	$\delta_x(Y)$ $\delta_z(Y)$	$\varepsilon_x(Y)$ $\varepsilon_y(Y)$ $\varepsilon_z(Y)$
Z-axis motion	$\delta_z(X)$	$\delta_x(X)$ $\delta_y(X)$	$\varepsilon_x(Z)$ $\varepsilon_x(Z)$ $\varepsilon_x(Z)$
Squareness errors		S_{xy} S_{yz} S_{zx}	

IOP Conf. Series: Journal of Physics: Conf. Series **1183** (2019) 012012 doi:10.1088/1742-6596/1183/1/012012

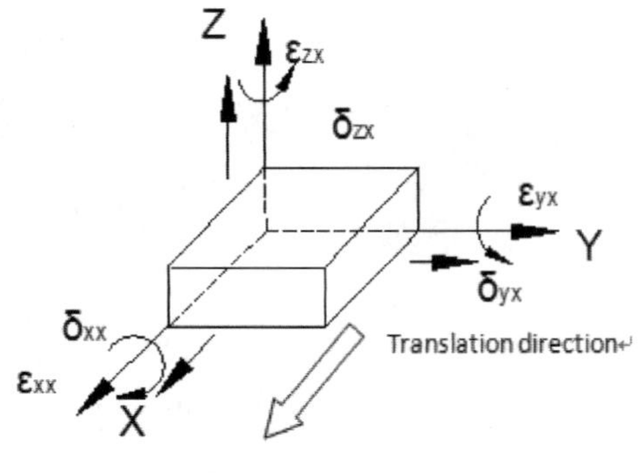

Figure 5. Six geomteric X-axis errors

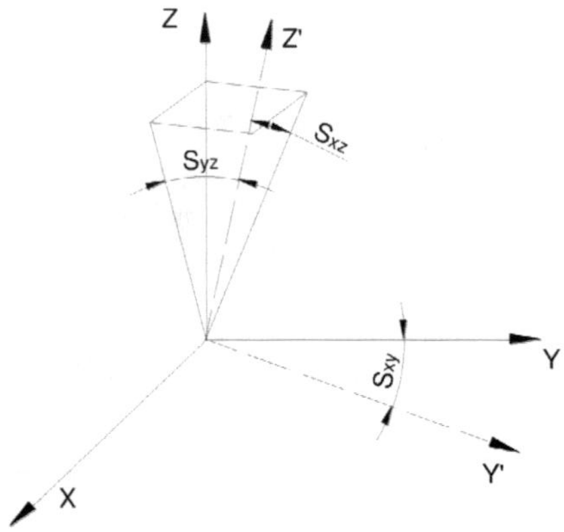

Figure 6. Space diagram of squareness errors

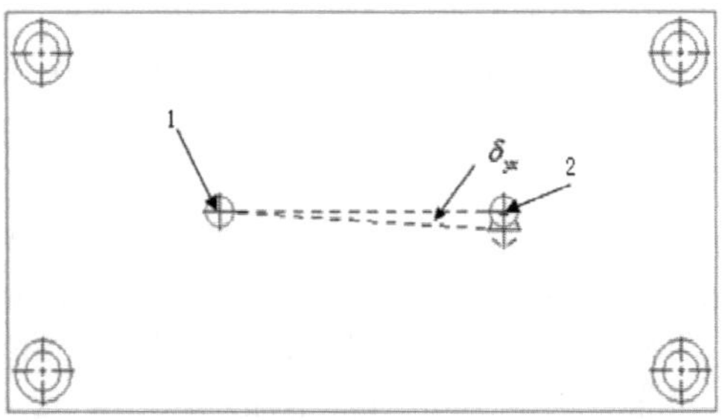

Figure 7. Straight line error detection

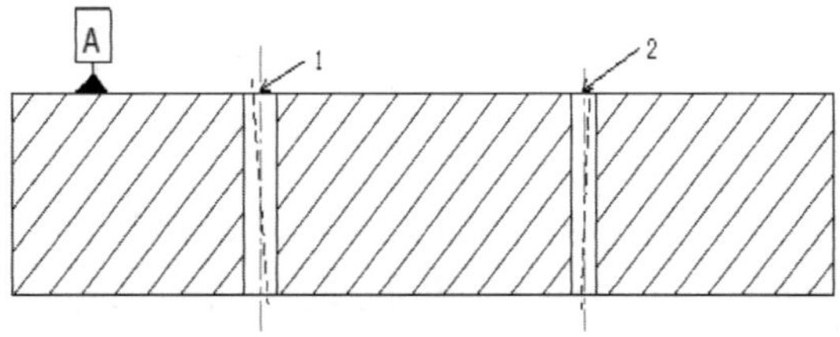

Figure 8. Squareness errors

3.1.2. Analysis of static accuracy of rotating axis

During the processing of the "S" Specimens, it is assumed that the deviation of the actual rotation axis of the C-axis in the X-axis follows the examples shown in figure 9.

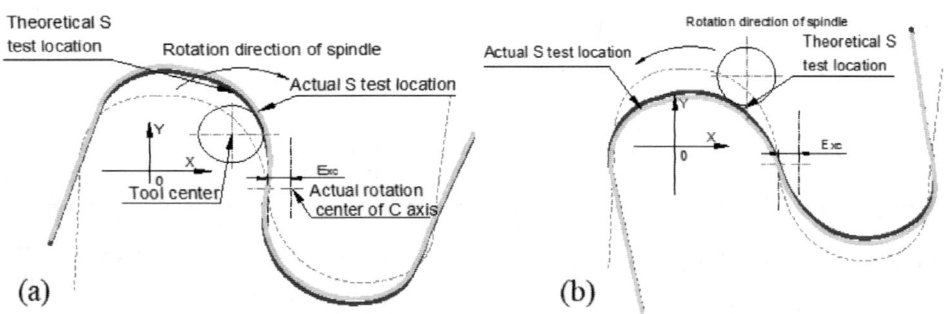

Figure 9. Contour machining (a) Inner contour machining and (b) External contour machining.

When machining the inner contour shown in figure 9a, the actual milling position is offset to the theoretical position clockwise. As a contrast,he actual milling position is offset to a certain value counterclockwise when machining the outer contour corresponding to this point as shown in figure 9b. This error results in the actual thickness of the point greater than the theoretical thickness of the point of the "S" Specimen. The actual distance between the A-A points of the corresponding points is measured (see figure 10a). According to the size of the difference the C-axis rotation axis offset in the X-direction can be determined. Similarly, the corresponding detection points can be set to detect the

IOP Conf. Series: Journal of Physics: Conf. Series **1183** (2019) 012012 doi:10.1088/1742-6596/1183/1/012012

offset in the Y-and Z-directions. Before machining the edge of the "S" specimen, the upper surface is machined with the same tool. If the error exists, the upper surface of the machined surface is inclined. After the "S" specimen is machined, the difference between the two-point B-B' of the upper surface is measured as shown in figure 10c).

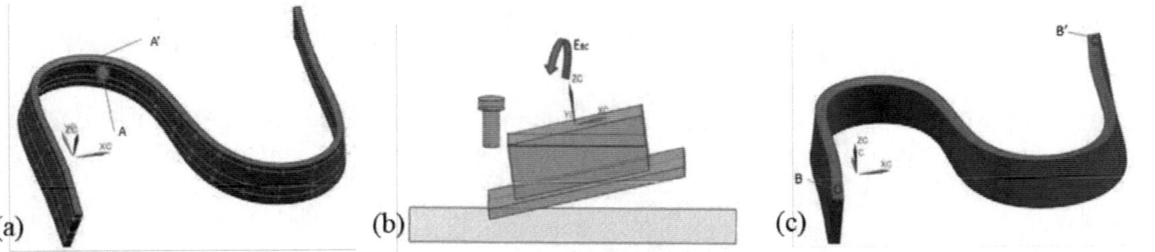

Figure 10. Contour machining (a) Schematic diagram (b) E_{BC} on the geometric error (c) Distribution of measuring points.

3.2. Analysis of dynamic error of machine tool on "S" specimen

Based on the multi-body system error analysis theory, the machining defect model of the machine tool is established. The expression of any point on the "S" surface is solved by using the method of superposition of tool center position and tool direction vector, and then the formula of the corresponding error of each machining point on the surface is derived.

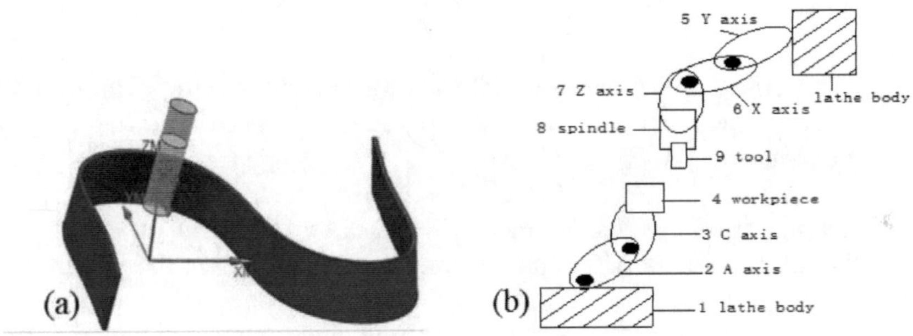

Figure 11. Tool and part in work-piece coordinate system. (a) Relationship between tool and part (b) RRTTT of MC.

The error position of each machining point can be divided into two branches: "machine tool-tool" branch and" machine tool-part" branch". A coordinate system established in each moving body based on the homogeneous coordinate transformation, the tool-center-point and tool-direction-vector can be described in the work-piece coordinate system through the motion transformation matrix error, which contains the actual tool path and the actual tool direction.

$$\{p_T\}_W = [B_W]^{-1}[B_T]\{p_T\}_T \tag{1}$$

$$\{v_T\}_W = [B_W]^{-1}[B_T]\{v_T\}_T \tag{2}$$

Where, $\{p_T\}_W$, $\{p_T\}_T$ respectively represent the coordinates of the position of the cutter head in the work-piece coordinate system and the cutter coordinate system. $[B_W]$ is the actual motion change matrix for the machine tool work-piece, $[B_T]$, i.e. the actual motion change matrix of machine tool cutting. As shown in figure 12, the "machine tool-tool" branch includes 1-5-6-7-8-9, while the "machine tool-work-piece" branch mainly includes 1-2-3-4.

IOP Conf. Series: Journal of Physics: Conf. Series **1183** (2019) 012012 doi:10.1088/1742-6596/1183/1/012012

The change matrix is as follows:

$$[B_W]=[S_12]_P[S_12]_{Pe}[S_12]_S[S_12]_{se}[S_23]_P[S_23]_{Pe}[S_23]_s[S_23]_{se}[S_34]_P \quad (3)$$
$$[S_34]_{Pe}[S_34]_s[S_34]_{se}$$
$$[B_T]=[S_{15}]_P[S_{15}]_{Pe}[S_{15}]_s[S_{15}]_{se}[S_{56}]_P[S_{56}]_{Pe}[S_{56}]_s[S_{56}]_{se}[S_{67}]_P[S_{67}]_{Pe} \quad (4)$$
$$[S_{67}]_s[S_{67}]_{se}[S_{78}]_P[S_{78}]_{Pe}[S_{78}]_s[S_{78}]_{se}[S_{89}]_P[S_{89}]_{Pe}[S_{89}]_s[S_{89}]_{se}$$

The relative position relationship between the tool and the work-piece in the work-piece coordinate system is shown in figure 11. The actual position point of the tool center point C from the tool end face D can be expressed as the follows:

$$[p_c]_W=[p_T]_W+D\times[v_T]_W \quad (5)$$

From equation (1) (2) (5) can obtain x_c, y_c, z_c. Suppose all the error parameters of x_c, y_c, z_c are zero. The coordinate of position point can be obtained with ideal condition. The error expression of the three directions is:

$$E_x = x_c - x_{ct} \quad (6)$$
$$E_y = y_c - y_{ct} \quad (7)$$
$$E_z = z_c - z_{ct} \quad (8)$$

The accuracy of "S" specimen is the compiled effects of all machine error. Through detection of the value of error variation between actual and measured values, the performance and accuracy of each moving axis can be evaluated.

3.3. Acquisition of characteristics of machining center based on "S" specimen

In order to obtain the information of MC running performance condition, the vibration, current, noise and temperature signals are used. The vibration of the machine tool will cause the amplitude of the spindle or worktable to exceed the allowable range, resulting in the deterioration of the machined surface of the part. In the machining process, the vibration will produce different degrees of noise, and the spindle current will also change at the same time. Vibration signals are collected by using DASP system (a vibration measuring system developed by Dongfang Vibration Company). A three dimensional piezoelectric acceleration sensor with bolt connection are mounted in the spindle and the work-piece (side of the "S" specimen base). A single direction acceleration sensor with electromagnetic block connection is installed in the A-and C-axis. The current signals of the linear axis and the rotating shaft are processed by a current sensor (CHK50R1). Each sensor is installed on the A-phase power line of each axis-inverter output. In addition, a RS-15A acoustic emission (AE) sensor equipped with 40dB amplifier collects the abnormal sound from spindle or produced by tool wear during machining. The AE sensor is installed on the headstock of the spindle box, shown as figure 12.

IOP Conf. Series: Journal of Physics: Conf. Series **1183** (2019) 012012 doi:10.1088/1742-6596/1183/1/012012

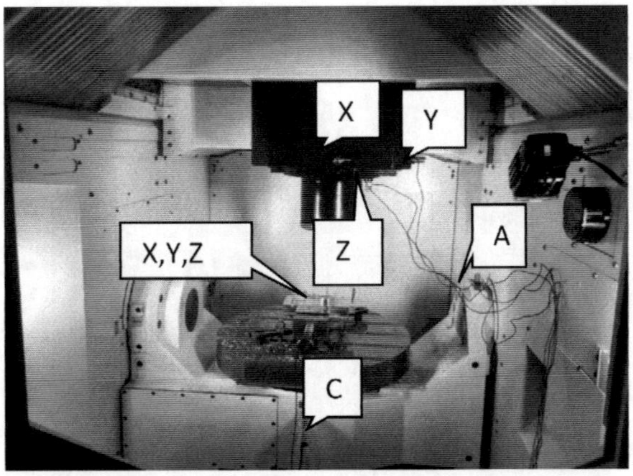

Figure 12. Sensors mounted in MAZARK MC

The evaluation model of the dynamic performance of the machining center is shown in figure 13. According to the curvature characteristics of the "S" specimen, the switching interval of opening and closing angle and the continuity of the surface, "S" specimen is divided into four regions. An acquisition and analysis system is developed to collect the signals during each milling layer processing based on LabVIEW. The detailed information of each cutting layer is collected, including current, vibration and noise signals. The processed "S" specimens were measured by a coordinate measuring machine and a scanning electron microscope respectively and the "S" specimen measuring data were imported into the database. The deviations of theoretical values and measured values of X-, Y- and Z-values at each point were analyzed. According to the acquisition time, the vibration, the current and the noise data of the corresponding machining section and the performance of the machine tool when machining the corresponding contour was evaluated.

The working flowchart as the follows:

- Obtained "S" specimen post-process instructions;
- Time discretization during the machining;
- While fine machining each layer of the "S" specimen, vibration, current, noise signal data are collected.
- Signal processed and in accordance with the different tester processing time sensitive information and corresponding processing area.
- Compared the standard qualified "S" specimen signals features with the features of the practical "S" specimen;
- Compare the characteristic signals, shapes and thresholds of the normal axes in the machining process are set up, and the surface morphology of the machined parts.
- Evaluation by using the practical information with the standard information.

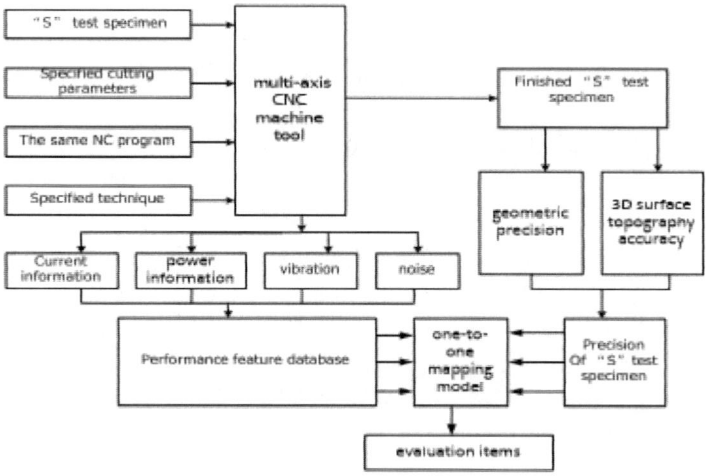

Figure 13. Running performance evaluation model for 5-axis MC based on "S" specimen.

4. Experimental verification

The evaluation model is used for a vertical five-axis MAZAK MC with a tilting rotary table, shown as figure 14.

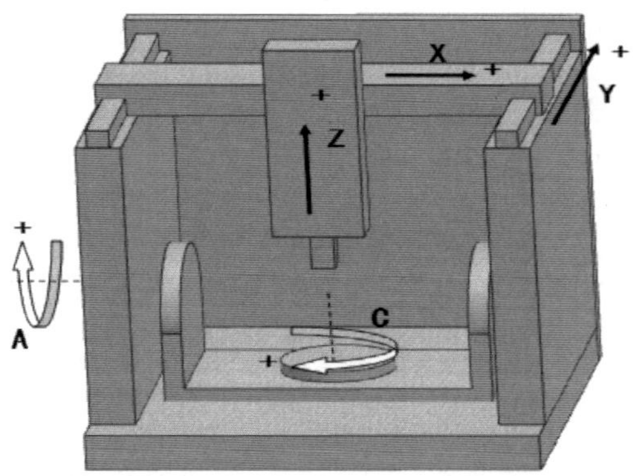

Figure 14. The vertical five-axis machining centre with a tilting rotary table

4.1. Analysis of three coordinate measurement results of the "S" specimen

The height of S edge strip is 20mm, and the offset value of three measuring lines to the upper surface are 2.5mm, 11.25mm, 16mm each.

The measuring lines are 1#line, 2#line and 3#line, seen as figure 15. The measurement results of the partial points are shown in Table 3, the analysis results could be seen clearly in figure 16-17.

Through the statistical analysis of the data points of three S curves, the vibration deviation of Y-direction was found to be larger, and the distance deviation at 16, 17 point is the biggest. This corresponds to the C -point of the part, the distance deviation of the 7, 8, 9 point is the second largest, and the point corresponds to the B-point of the part surface.

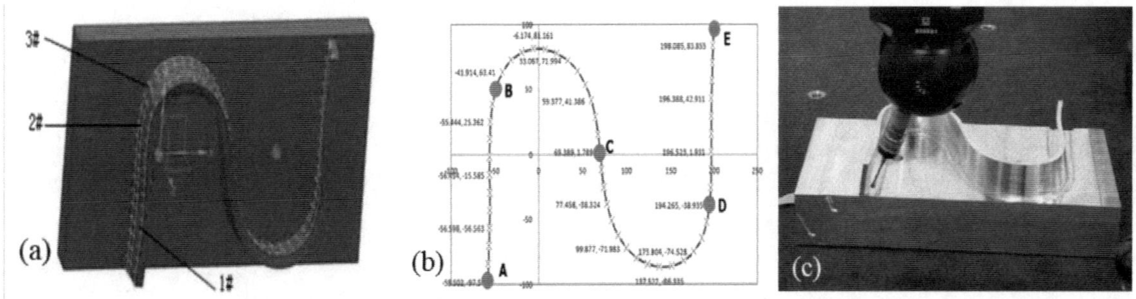

Figure 15. "S" specimen measuring with CMM (a) three measuring lines (b) measurement curve (c) measurement site.

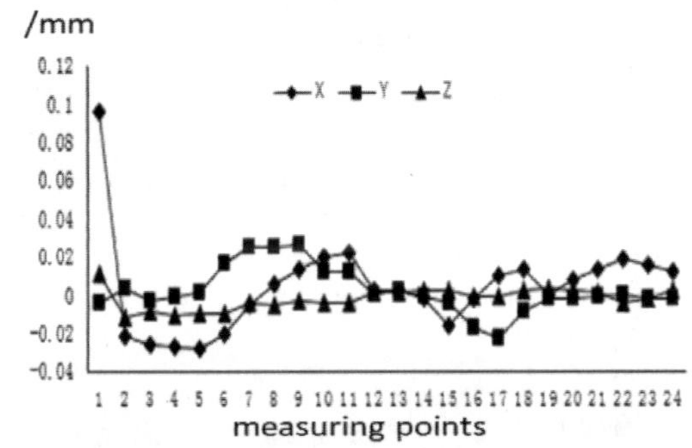

Figure 16. Distance deviation in X-, Y-, and Z- directions along 3# line.

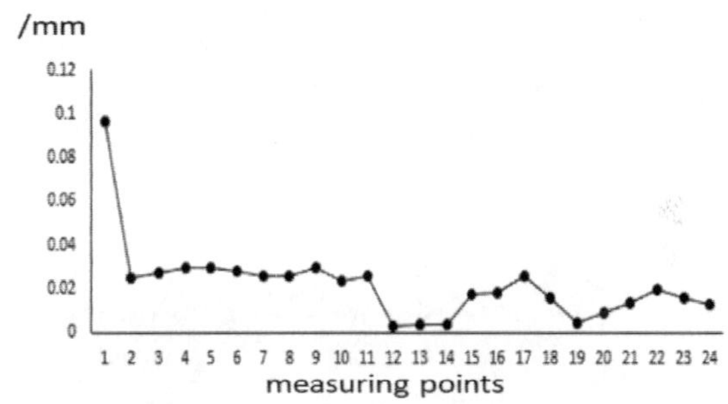

Figure17. Total distance deviation along 3# line.

Table 2. Geometrical values of surface points of the "S" specimen

Theoretical value (mm)			Actual measured value (mm)			Deviation value (mm)		
X	Y	Z	X'	Y'	Z'	ΔX	ΔY	ΔZ
94.098	49.0568	17.1305	94.2645	49.0473	17.1438	0.1665	-0.0095	0.0133
92.6807	35.2609	17.0129	92.6705	35.2638	17.007	-0.0103	0.0029	-0.0059
91.5905	21.4388	16.9348	91.5816	21.4397	16.931	-0.0089	0.001	-0.0037
91.6503	7.5722	16.9551	91.638	7.5691	16.9513	-0.0123	-0.003	-0.0037
95.237	48.9741	8.4553	95.3607	48.9677	8.4645	0.1237	-0.0064	0.0093
94.173	34.8289	8.395	94.159	34.8344	8.3855	-0.014	0.0054	-0.0095
93.3683	20.6687	8.3712	93.3578	20.6663	8.3644	-0.0104	-0.0024	-0.0068
93.4246	6.4847	8.3841	93.4103	6.4828	8.3772	-0.0143	-0.002	-0.0069
93.5962	-7.6986	8.415	93.5794	-7.7004	8.4108	-0.0168	-0.0018	-0.0041
92.1615	-21.7817	8.4486	92.1489	-21.7681	8.4403	-0.0126	0.0135	-0.0083
95.8564	48.9307	3.746	95.9526	48.9266	3.7566	0.0962	-0.0041	0.0106
94.9838	34.5958	3.7168	94.9618	34.599	3.7049	-0.022	0.0032	-0.0118
94.3338	20.2507	3.7222	94.308	20.2479	3.7139	-0.0258	-0.0028	-0.0083
94.3874	5.89	3.7311	94.36	5.8885	3.7204	-0.0274	-0.0015	-0.0106
94.4243	-8.4701	3.7422	94.3965	-8.4691	3.7326	-0.0277	0.0009	-0.0097

IOP Conf. Series: Journal of Physics: Conf. Series **1183** (2019) 012012 doi:10.1088/1742-6596/1183/1/012012

4.2. Analysis of the sensor signals during "S" specimen machining

The vibration signals from the part during machined can be obtained as shown in figure 18. Vibration signals from the spindle are shown in figure 19. Analysis of the relationship between signals obtained during the actual processing and the corresponding geometrical measuring point's data (distance deviations) provide a way of linking the on-line sensor data to expected geometrical deviations.

Sensor signals processed by using db4 wavelet, time domain and frequency domain are listed in table 3.

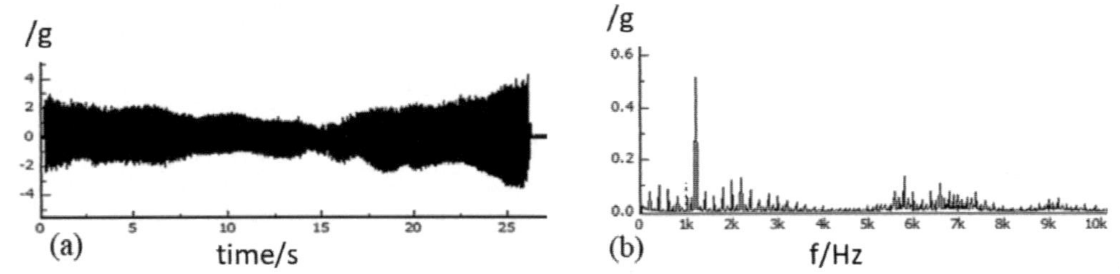

Figure 18. The vibration signal in X-direction from part (a) Original signal (b) frequency domain.

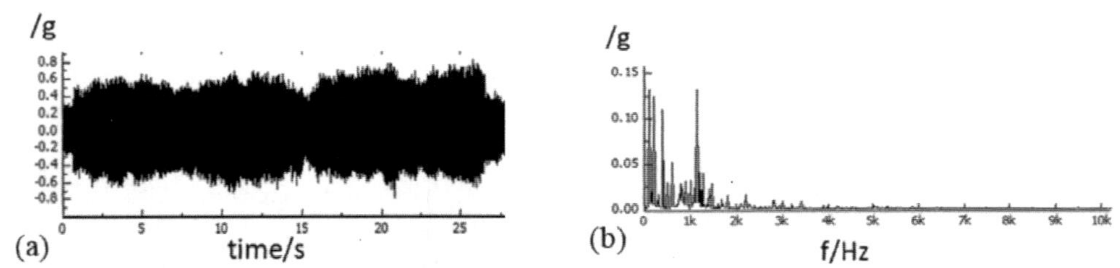

Figure 19. The vibration signal in X -direction from spindle (a) Original signal (b) frequency domain.

Table 3. Main sensor signals from the monitoring system

Mean value (g)	margin indicators	kurtosis	Mean frequency (Hz)	Relative energy of first frequency band
0.165287	8.224465	4.253465	630.610274	0.333490
0.203196	4.706014	3.436329	652.885428	0.331180
0.203516	4.356667	3.214722	620.288575	0.323180
0.213184	4.296957	2.999977	677.691592	0.337285
0.241505	4.919377	2.919862	582.241655	0.329824
0.292100	4.866367	2.904250	625.132254	0.323451
…	…	…	…	…
0.221508	5.384831	2.852440	557.001919	0.323676
0.305253	5.326490	2.982224	678.951130	0.335667
0.374143	5.306549	2.965261	576.347319	0.308494
0.394331	5.052041	3.060290	670.357880	0.342776
0.280102	5.107896	3.013980	662.324249	0.341480
0.264127	4.152854	2.896036	809.493795	0.376218
0.283333	4.600816	2.949488	760.479086	0.365216
0.299129	5.026934	3.051128	801.873472	0.375502
0.262652	8.668569	3.935178	719.692620	0.363176
0.013957	2.739557	2.654404	417.106758	0.252520

IOP Conf. Series: Journal of Physics: Conf. Series **1183** (2019) 012012 doi:10.1088/1742-6596/1183/1/012012

The surface finish value of the "S" specimen is obtained by using the SEM. Here, the points on the "S" specimen with especially sensitive to large surface deviations (vibration patterns, waviness) can be identified (see table 4) and linked to predicted geometrical deviations of the final machined part, therefore establishing a normal status pattern of the sensor signals for a geometrically measured and qualified "S".

If the monitored signals are deviating from the reference signal significantly, a warning signal could be sent to the operator of the MC.

Table 4. Mapping relationship between the different features

Point of "S" Specimen	Point coordinate data(mm)	Signal from monitoring system	Surface finish
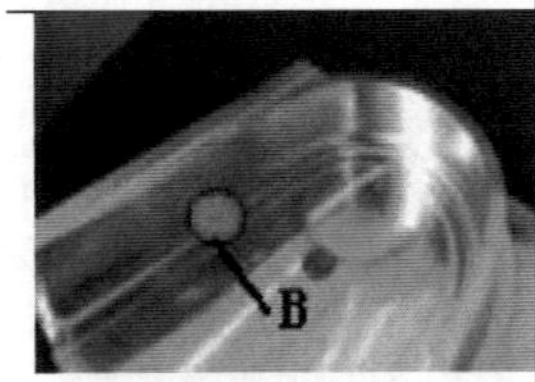	-30.2150, 9.3818, 4.7011	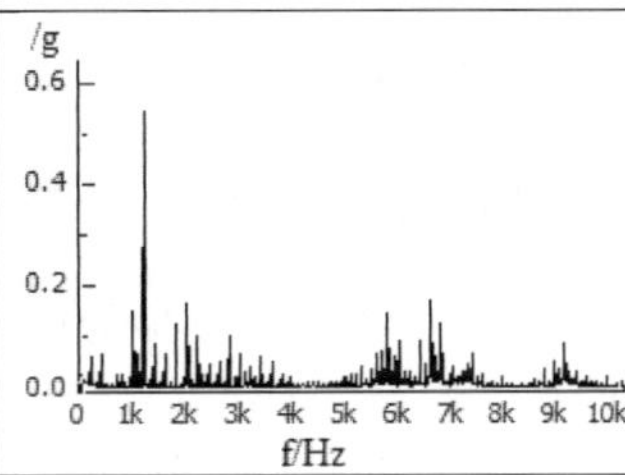	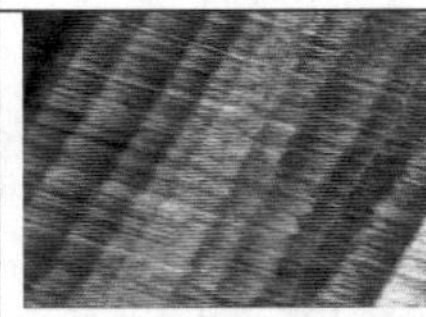
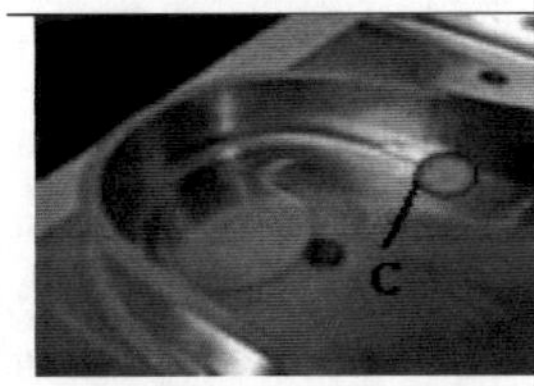	29.4213, 26.3824, 4.5363	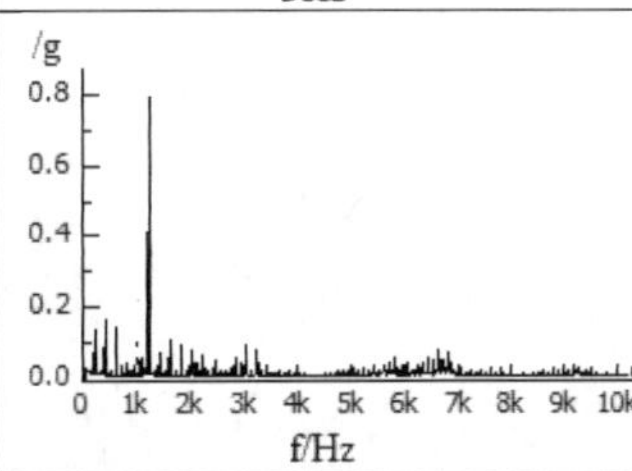	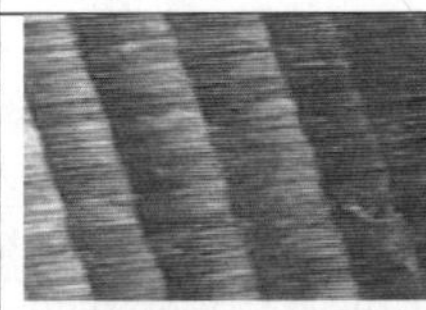
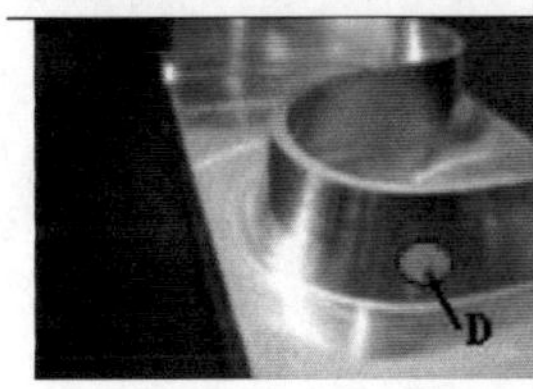	92.6866, -22.6927, 3.7564	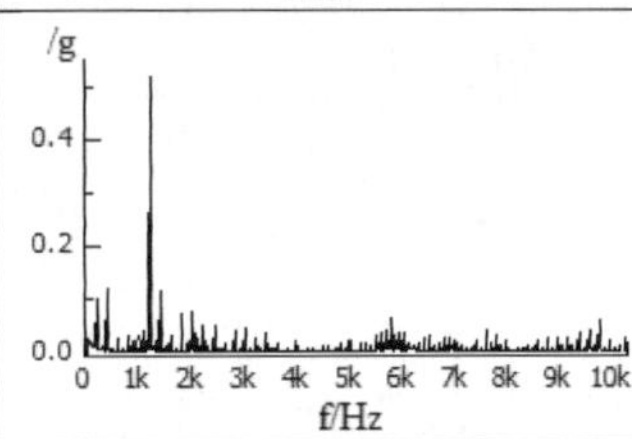	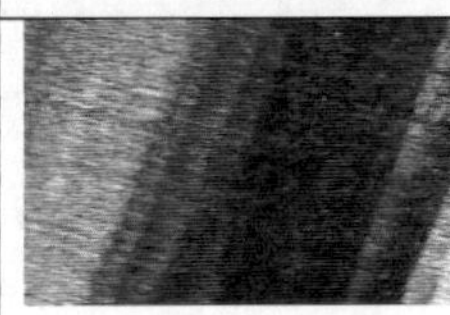

5. Conclusions and Future work

The paper analyzes the advantages and disadvantages of the current five-axis MC inspection and testing parts, the structural characteristics of S-shaped specimen. The S-shaped specimen can reflect the kinematic dynamic performance, machining precision, machining condition and structure rigidity and vibration of the Machine tool. The main conclusions and future issues from this study are:

- A method for evaluating the operation state of 5-axis MC based on S-specimen is proposed.
- A machine tool running condition evaluation system based on S specimen is developed. The data acquisition system can collect dynamic signals included vibration, current, temperature and connect to expected geometrical deviations including surface finish (waviness, chatter, and roughness).
- Furthermore, in the future, there are still a lot to do, such as how to judge the running performance of the 5-axis MC more quickly and with high precision. Additional experiments are required to build a database where sensor data can be robustly connected to geometrical measured deviations, surface finish and control algorithms to minimize the deviations and errors of the produced parts. .

- Another future point is to more effectively characterize the surface finish using in-line- or laboratory surface profilers and ISO 4287 and ISO 25178-2 profile- and areal parameters.

6. Acknowledgement

Authors are grateful to the anonymous reviewers and the editor for their critical and constructive review of the manuscript. This research is supported by National Natural Science Foundation of China (Grant No.51575055) and National Science and Technology Major Project of China (Grant No.2015ZX04001002).

7. References

[1] Qiu J, Wang G M and Gen R P 2013 Roundness errors diagnoses of 5-axis horizontal machining center based on DBB Manufacturing Technology & Machine Tool .08 39-42+46.
[2] Gao X F, Liu Ch Sh, Li Y, Lin J F, Xu J C and Hua Ch Ll 2011 Positioning errors detection and identification for A/C axes bi-rotary milling head based in laser interferometer Machinery Design & Manufacture 12 212-14.
[3] S. Weikert 2004 R-test, a new device for accuracy measurements on five axis machine tools CIRP Annals 53(1) 429-32.
[4] NAS979 uniform cutting test. NAS series, metal cutting equipment specifications.1969:34-7
[5] Song Z Y and Cui Y W 2007 S-shape detection test piece and a detection method for detection the precision of the numerical control milling machine CN 101000285A.
[6] Chen D J, Dong L H, Gao X, Wang X F and Fan J W 2016 Error tracing of the complex CNC machine tools based on “S” shaped work piece Journal of Mechanical Engineering 52(11) 155-64.
[7] Wang W,Jiang Zh, Li Q C and Tao W J 2015 A new test part to identify performance of five-axis machine tool-part validation of S part Int. J. Adv. Manuf. Technol .79 739-56
[8] Jiang Zh, Ding J X, Song Zh Y, Du L and Wang W 2016 Modeling and simulation of surface morphology abnormality of “S” test piece machined by five-axis CNC machine tool Int. J. Adv. Manuf. Technol. 85 2745-59.
[9] Qiu J 2014 S form specimen cutting dynamic performance testing and its key technologies China Mechanical Engineering 25(12)1600-04, 1629
[10] Soichi I, Masahiro S, Atsushi M and Tetsuya M 2010 Machining tests to identify kinematic errors on five-axis machine tools Precision Engineering 34 387-98
[11] G H J Florussen, HA M.Spaan and T.M.Spaan-Burke 2016 Assessing the accuracy of five axis machines by comparing machine measurement data with test work piece deviations Procedia Manufacturing 625-32
[12] ISO 10791-7 2014 Test conditions for machining centers, part 7.Accuracy of finished test pieces. ISO (International Organization for Standardization)
[13] ISO 230-7 2006 test code for machine tools, part 7 Geometric accuracy of axes of rotation
[14] Wang H J, Xu X L and Wan P 2014 Rotor System Fault Diagnosis Based on Orbit Manifold Topological Journal of Mechanical Engineering 20(5) 95-101
[15] Wan P, Wang H J and Xu X L 2012 Study of fault diagnosis model based on local tangent space alignment and support vector machine Chinese Journal of Scientific Instrument 33(12) 2789-95

Dynamic characteristics of the annular rotor components in a rotor-bearing system under high frequency stress conditions

Chenglin Wang[1], Huating Chen[1], Weichen Sun[2] and H J Wang[3,4]
[1]Beijing Wuzi University, China.
[2]Beijing Institute of Technology, China.
[3]School of Mechanical and Electrical Engineering, Beijing Information Science and Technology University, China.
[4]Key Laboratory of Modern Measurement and Control Technology, Ministry of Education, Beijing Information Science and Technology University, China.

E-mail: wangchenglin6688@126.com

Abstract. The dynamic characteristics of rotor components can directly affect the stability of a rotor-bearing system. Under certain conditions, such as high frequency stress, the dynamic performance of the rotor components is of particular importance. In this paper, a typical annular rotor is studied, and a testing system presented based on the base excitation resonance mass (BERM) method in order to measure key parameters, such as the radial and axial stiffness coefficients, damping coefficient, and loss factor. Based on this system, the dynamic characteristics of typical rotor components under radial and axial sine excitations below 2000 Hz were tested and analyzed. A predictive model of the characteristic parameters was then built based on the grey prediction method, and coupling between the radial and axial dimensions was evaluated. Further, the coupling mechanisms and relationships were analyzed, and the mechanism of interaction between microstructure features under high frequency stress was determined and combined with the changes to the internal microstructure of the rotor components to build a complete picture. Based on the experimental testing and theoretical results, our work suggests a strong correlation between the rotor stiffness coefficient and frequency of use, and thus provides empirical evidence and a reference for the future optimization of the dynamic characteristics of rotor components, in particular for the design of parameters for rotor components the experience high frequency stress conditions on the bearings.

Keywords: Dynamic characteristics, High frequency bearing characteristics, rotor components, grey prediction.

1. Introduction

With the rapid development of modern engineering and technology, mechanical devices are becoming more widely used and often operate under high frequency stress conditions, in particular, with the performance of rotating machinery being continuously improved by increasing rotational speeds. Rotor components are the most critical components of rotating machinery, and their dynamic characteristics directly affect the stability of the system, especially under certain loading conditions, including high frequencies and high acceleration. Moreover, dynamic performance is particularly important, as it will directly affect the potential usage of the system. Therefore, studying the mechanical properties of rotor components under high acceleration bearing conditions is of great importance [1-5].

IOP Conf. Series: Journal of Physics: Conf. Series **1183** (2019) 012013 doi:10.1088/1742-6596/1183/1/012013

2. Design of the testing system

2.1. System design principles

New materials are continuously being using in high frequency bearing applications. However, the quantitative relationships between the characteristic parameters, such as the stiffness coefficient, damping coefficient, loss factor, and geometry, and the operational parameters have not been fully understood so far [6-7]. There are numerous methods for testing the dynamic characteristics of rotors, for which the base excitation- resonance mass (BERM) method is a useful one. With the BERM method, the test object is placed between the excitation source and forced vibration body, and the dynamic characteristics of the test object are calculated and analyzed in order to acquire the characteristic dynamic parameters, including the stiffness coefficient and damping coefficient, by measuring the extracted feature information about the excitation source and forced vibration body, respectively. The dynamic model is based on the testing principles shown in figure 1. A shaking table is used as the source of excitation, and the test object can be simplified as the connection mechanism with stiffness coefficient and damping coefficient, for which the spring and damper are characterized by the eigenvalues (figure 1).

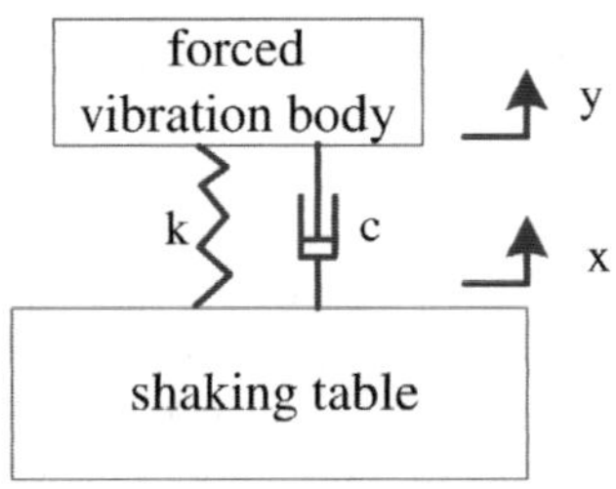

Figure 1. Dynamic model of BERM method.

Assuming the mass of the forced vibration body is m, the equation of motion can be written as

$$m\ddot{y} + c(\dot{y} - \dot{x}) + k(y - x) = 0 \tag{1}$$

The excitation form of the shaking table can be represented as

$$x = x_0 e^{i\omega t} \tag{2}$$

The response of the forced vibration body can be described as

$$y = y_0 e^{i(\omega t - \varphi)} \tag{3}$$

Where, ω is the vibration frequency.

Combining the above equations, the elastic coefficient *k* of the spring can defined as

$$k = \frac{m\omega^2(\alpha^2 - \alpha\cos\varphi)}{\alpha^2 - 2\alpha\cos\varphi + 1} \tag{4}$$

Where, φ is the phase difference.

The damping coefficient *c* of the tested object can be given by

$$c = \frac{m\omega\alpha\sin\varphi}{\alpha^2 - 2\alpha\cos\varphi + 1} \tag{5}$$

IOP Conf. Series: Journal of Physics: Conf. Series **1183** (2019) 012013 doi:10.1088/1742-6596/1183/1/012013

Where the amplitude ratio is

$$\alpha = \frac{y_0}{x_0} = \frac{\ddot{y}_0}{\ddot{x}_0} \tag{6}$$

and the loss factor β is

$$\beta = \frac{c\omega}{k} \quad . \tag{7}$$

Therefore, as long as the excitation amplitude x_0 can be measured, the amplitude ratio x_0 of the forced vibration body to the phase difference between them, as well as the stiffness, damping coefficient, and loss factor value of the test object can be acquired.

2.2. Structural design of the testing system

Rotor components have a complex structure. In order to reflect their main dynamic characteristics, in this paper we selected widely used annular components as the test objects, and the form of component fit was selected as the most widely used shaft-hole fit, between the annular components and the excitation source and forced vibration body. To simulate the actual working environment, concentric mounting holes were set at the front and back supporting ends on the excitation source (shaking table) for installing annular test components, and the forced vibration body was machined into a shaft component form, and installed within the components.

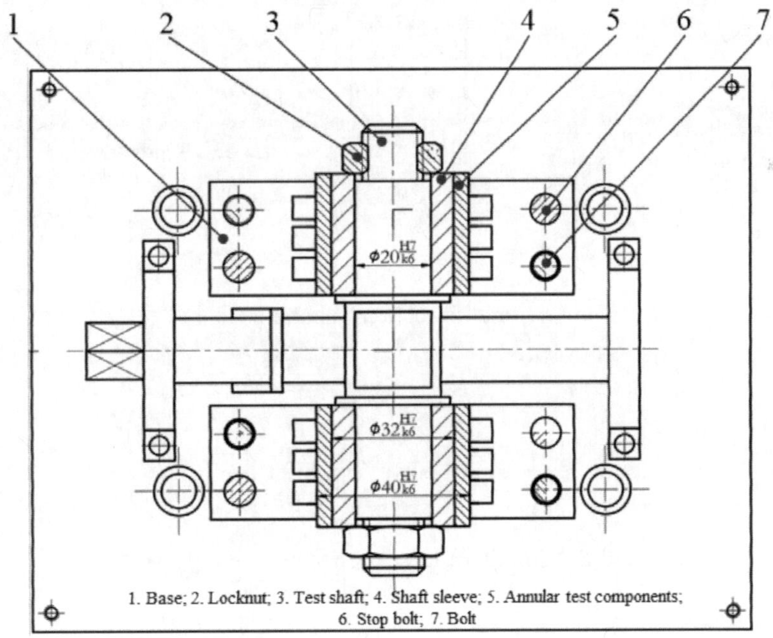

Figure 2. Mechanical drawing of the testing system.

To improve the system generality for testing the dynamic characteristics of annular components of different thicknesses, the fit between the annular components and forced vibration body can be realized by configuring appropriate shaft sleeves. Thus, we can achieve the rapid replacement of test objects thereby reducing the number of components that need to be replaced in the testing system. Figure 3 shows the assembly relationship diagram of the test objects.

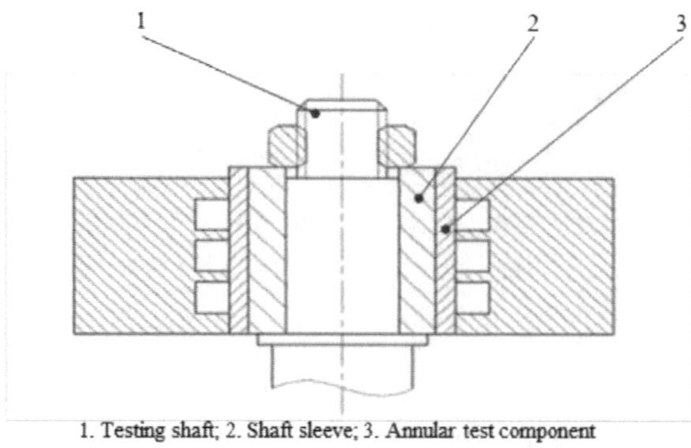

Figure 3. Drawing of the assembly relationship of the test objects.

2.3. Design of testing scheme

For current applications, rotor components are mainly composed of metallic and polymeric materials. In this study, both a metallic and polymer material were selected, aluminum Al6061 and nylon, respectively. The geometric parameters of the annular test components included the outer diameter (mm), inner diameter (mm), and length (mm), and the layout of the experimental testing system is shown in figure 4.

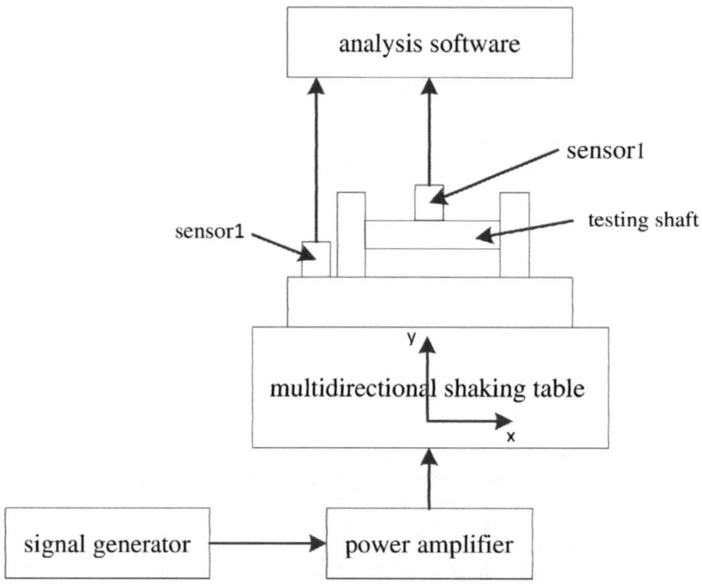

Figure 4. Schematic of the experimental testing system.

Three-dimensional acceleration sensors were installed on the shaking table and test object, and the acceleration signals were collected using a data acquisition instrument. The data was transferred to the analysis software, which reads the amplitude and phase information from the signals, and the dynamic characteristics can then be calculated. By controlling the feature parameters such as the excitation source, stress value, and excitation amplitude, the impact of frequency on the dynamic characteristics of the rotor components can be studied. Based on the demands for our desired usage, this paper mainly focuses on measuring stiffness coefficients, but will also analyze the relationship between the stiffness coefficient and frequency of use.

IOP Conf. Series: Journal of Physics: Conf. Series **1183** (2019) 012013 doi:10.1088/1742-6596/1183/1/012013

The temperature during testing was 5°C-10°C, and the model number and performance parameters of the instruments used in the experiment are listed in Table 1.

Table 1. Model number and performance parameters of instruments used in the experiment.

Instrument	Performance parameters	Model number
Shaking table	Maximum working frequency = 2000 Hz Three-dimensional simultaneous vibration sinusoidal thrust = 9.8 kN	Suzhou Sushi Testing Instrument Co., Ltd. MAV-1000-4H
Acceleration sensors	Three-dimensional simultaneous measurement Frequency range = 1-4000 Hz	China orient institute of noise & vibration INV9832
Data acquisition instrument	Supporting 8 channel signal input Supporting signal input by BNC connector	China orient institute of noise & vibration INV3062
Analysis software	Possessing basic functions including oscillographic sampling, time domain analysis, auto-regression spectrum analysis, functional analysis	China orient institute of noise & vibration DASO-V10, Featured Version

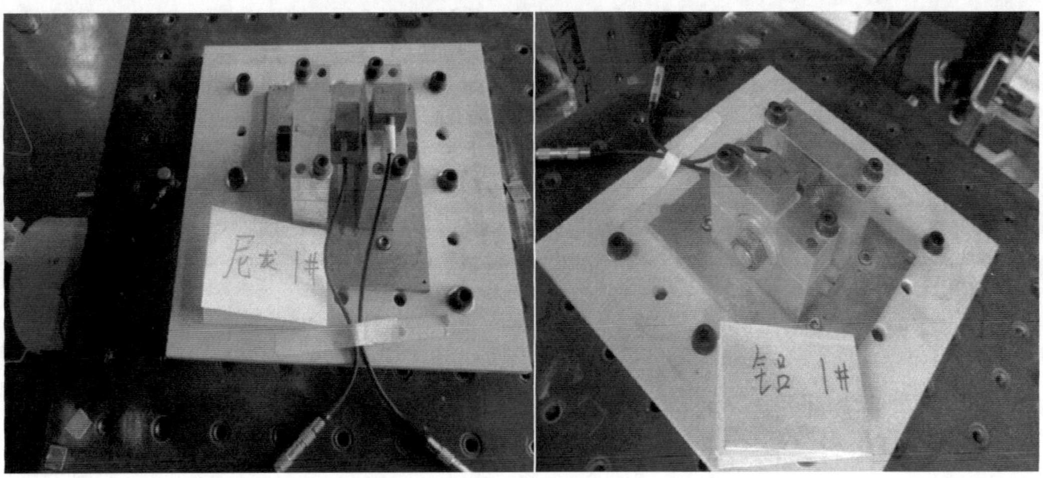

Figure 5. Spot map of the experimental test.

IOP Conf. Series: Journal of Physics: Conf. Series **1183** (2019) 012013 doi:10.1088/1742-6596/1183/1/012013

3. Analysis of experimental results

Due to the application of high-frequency stress on the rotor-bearing system, there is some level of uncertainty in the experimental results at certain frequencies, due to the influence of the dynamic characteristics of the testing system itself. These uncertainties mainly manifest in the discrete nature and relatively strong distribution inhomogeneity of the experimental data, which greatly influences the overall test results. For convenient statistical analysis, the interval frequency of adjacent test points within the testing range was set to 100 Hz, and the testing frequency values were set below 2000 Hz. In the testing process, abnormal data were eliminated, and a typical sample data set was selected for closer analysis. For the experiment, the test object was an aluminum Al6061 specimen of 40 mm external diameter and 2.5 mm thickness, and a hard nylon sample of the same measurements. The results are shown in Tables 2 and 3.

Table 2. Stiffness test results of Al6061 aluminum alloy component.

Axial data of unidirectional testing	Frequency (Hz)	700	1100	1400	1700
	Stiffness (N/m)	1008030.408	2652754.724	5384916.58	19588536.06
Axial data of two-dimensional coupling testing	Frequency (Hz)	400	600	1100	1700
	Stiffness (N/m)	299542.3509	1433397.768	3042655.44	7140040.754
Radial data of unidirectional testing	Frequency (Hz)	500	700	1000	1500
	Stiffness (N/m)	576793.1802	1312774.967	2301878.82	11317649.19
Radial data of two-dimensional coupling testing	Frequency (Hz)	400	700	1200	1600
	Stiffness (N/m)	293135.2191	624385.4955	1370196.88	14436590.1

3.1. Predictive analysis of the stiffness coefficient based on the grey prediction method

Considering the discrete nature and distribution inhomogeneity of the experimental data, the sample data must be processed to acquire characteristic parameter values under loading at different frequencies. Combined with the data distribution characteristics, the grey prediction method can be used to analyze the stiffness data [8].

Table 3. Stiffness test results of hard nylon component.

Axial data of unidirectional testing	Frequency (Hz)	1300	1400	1700	2000
	Stiffness (N/m)	971614.709	1042198.05	1213740.18	3551248.47
Axial data of two-dimensional coupling testing	Frequency (Hz)	700	1100	1800	1900
	Stiffness (N/m)	454704.895	2934442.573	4108264.248	4327876.048
Radial data of unidirectional testing	Frequency (Hz)	500	700	1000	1700
	Stiffness (N/m)	569118.3472	1393918.973	2260463.347	6341113.142
Radial data of two-dimensional coupling testing	Frequency (Hz)	400	900	1500	1800
	Stiffness (N/m)	287127.4225	666801.7568	886089.8392	4310155.261

3.2. Establishment of the grey prediction model

In order to reduce the influence of the high growth sequence on the fitting accuracy, the original sequence data are subjected to a logarithmic transformation. Here, the original sequence data are represented by $X^{(00)}(K_i)$, and the transformed logarithmic sequence data is represented by $X^{(01)}(K_i)$, namely $X^{(01)}(K_i) = \ln(X^{(00)}(K_i))$.

Assuming, $X^{(0)}(K_i) = \{X^{(0)}(K_1), X^{(0)}(K_2), \dots X^{(0)}(K_n)\}$, if the internal $\Delta K_i = K_i - K_{i-1}$, *i=2*, *3…n*, is not a constant, then $X^{(0)}(K_i)$ is a non-equidistant sequence.

Assuming $X^{(1)}(K_i) = \{X^{(1)}(K_1), X^{(1)}(K_2), \dots X^{(1)}(K_n)\}$, if $X^{(1)}(K_i) = \sum_{j=1}^{i} X^{(01)}(K_j)\,\Delta K_j$, *i=1*, *2…n*, then $X^{(1)}(K_i)$ is the first-order accumulated generating (1-AGO) sequence of the non-equidistant sequence $X^{(0)}(K_i)$.

For the 1-AGO sequence, the albinism differential equation can be built as follows:

$$\frac{dx_1^{(1)}(t)}{dt} + a x_1^{(1)}(t) = u \quad \text{i=1,2,…,n} \tag{8}$$

If we stipulate $X^{(1)}(K_1) = X^{(0)}(K_1)$ when t=K_1, then the response function is

$$\widehat{x_1^{(1)}}(K_i) = \left[X^{(0)}(K_1) - \frac{u}{a}\right] e^{-a(K_i - K_1)} + \frac{u}{a} \text{i=1,2,…,n} \tag{9}$$

After reduction, the expression of the model is

$$\widehat{x_1{}^{(0)}}(K_{i+1}) = \frac{1}{\Delta K_{i+1}}(1 - e^{a\Delta K_{i+1}})\left[X^{(0)}(K_1) - \frac{u}{a}\right]e^{-a(K_{i+1}-K_1)} \tag{10}$$

The summary parameters of the above model, a and u, can be obtained through $[a,u]^T = (B^T B)^{-1}B^T Y_N$, where

$$B = \begin{bmatrix} -\boldsymbol{Z}^{(1)}(\boldsymbol{K}_2)\mathbf{1} \\ -\boldsymbol{Z}^{(1)}(\boldsymbol{K}_3)\mathbf{1} \\ \vdots \\ -\boldsymbol{Z}^{(1)}(\boldsymbol{K}_n)\mathbf{1} \end{bmatrix}$$

Where, $Z^{(1)}(K_i)$ is the background value of $X^{(1)}(t)$ in the interval $[K_{i-1}, K_i]$. The background value can be calculated by the two-point smoothing expression of the traditional GM（1, 1）model as follows:

$$Z^{(1)}(K_{i+1}) = \frac{X^{(1)}(K_{i+1})+X^{(1)}(K_i)}{2} \tag{11}$$

$$Y_N = \left[X^{(0)}(K_2), X^{(0)}(K_3), \dots X^{(0)}(K_n)\right]^T \tag{12}$$

3.3. Numerical analysis of the stiffness coefficients based on the grey prediction method
Based on the established grey prediction model, the predicted stiffness coefficients can be calculated in MATLAB, and the results are shown in Tables 4 and 5.

Table 4. Prediction values of stiffness coefficients of Al6061.

Prediction values of unidirectional axial stiffness	Frequency (Hz)	700	1100	1400	1700
	Stiffness (N/m)	1008030.408	2355132.003	6681198.181	17347219.72
Prediction values of coupling axial stiffness	Frequency (Hz)	400	600	1100	1700
	Stiffness (N/m)	299542.3503	1520750.969	2750937.377	7327906.453
Prediction values of unidirectional radial stiffness	Frequency (Hz)	500	700	1000	1500
	Stiffness (N/m)	576793.1794	1190456.029	2653104.184	10561344.75

Prediction values of coupling radial stiffness	Frequency (Hz)	400	700	1200	1600
	Stiffness (N/m)	293135.219	489507.4992	2041140.825	12115988.25

Table 5. Prediction values of stiffness coefficients of hard nylon.

Prediction values of unidirectional axial stiffness	Frequency (Hz)	1300	1400	1700	2000
	Stiffness (N/m)	971614.709	912551.681	1501537.288	3275234.409
Prediction values of coupling axial stiffness	Frequency (Hz)	700	1100	1800	1900
	Stiffness (N/m)	454704.895	3027960.958	3810610.612	4512220.767
Prediction values of unidirectional radial stiffness	Frequency (Hz)	500	700	1000	1700
	Stiffness (N/m)	569118.3472	1394295.74	2244963.902	6189717.465
Prediction values of coupling radial stiffness	Frequency (Hz)	400	900	1500	1800
	Stiffness (N/m)	287127.4225	523451.678	1447574.564	3512013.227

The precision of the algorithm was tested using the parameters: model precision p and posterior error ratio C. From the grey prediction model of Al6061, $p = 99.2\%$ and $C = 0.25$ for axial data with unidirectional testing. According to the experimental results, the model precision is at the first level, wherein $p = 99.7\%$ and $C = 0.049$ for axial data with two-dimensional coupling. In the grey prediction model of hard nylon, $p = 99.9\%$ and $C = 0.006$ for radial data with unidirectional testing, and the model precision is of the first level according to the experimental results was $p = 98.3\%$ and $C = 0.301$ for radial data using two-dimensional coupling testing. Again, the model precision was first level.

3.4. Analysis of stiffness variation under two-dimensional coupling vibration.

Under different conditions of high frequency stress loading, the dynamic characteristics of the toroidal rotor components, especially the stiffness coefficient will change. The key factors include the size of the stress value, the direction and frequency of stress, etc., especially after high-frequency fatigue loading, the grain structure of the rotor components will change. This article focused comparing the grey prediction results for unidirectional vibration with those of two-dimensional coupling vibrations, the influence of two-dimensional coupling vibration on stiffness can be determined.

IOP Conf. Series: Journal of Physics: Conf. Series **1183** (2019) 012013 doi:10.1088/1742-6596/1183/1/012013

The grey prediction model was fitted by the cftool toolbox in Matlab. By analyzing the stiffness formula, an exponential relationship was found between stiffness and frequency. Hence, $y = ax^b + c$ was selected as the fitting equation. Thus, the fitting results for the aluminum alloy annular components are based on the radial uncoupling stiffness equation $k = 1.26\times10^{-5}w^{3.747}+4.928\times10^5$, radial coupling stiffness equation $k = 3.145\times10^{-13}w^{6.113}$, axial uncoupling stiffness equation $k = 1.156\times10^{-8}w^{4.694}+3.424\times10^5$, and axial coupling stiffness equation $k = 0.2525w^{2.302}+4\times10^5$.

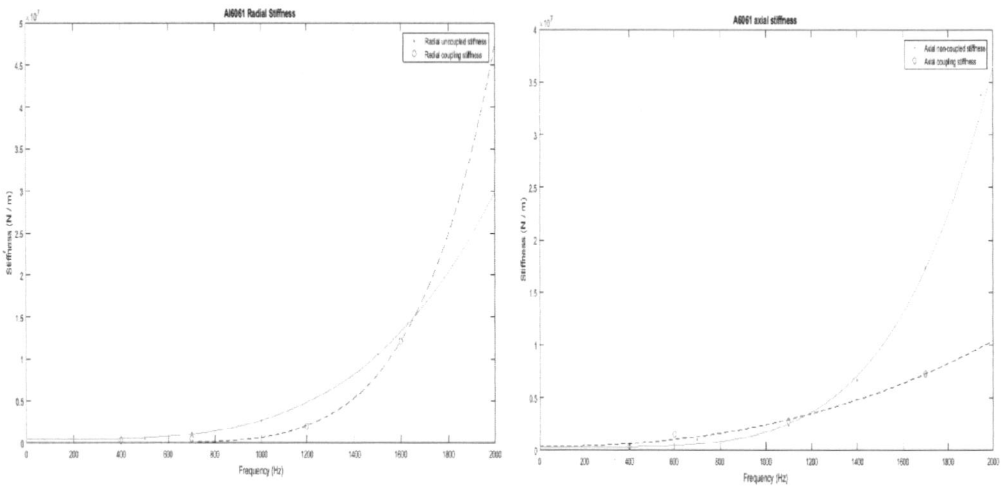

Figure 6. Fitted stiffness curves of the aluminum alloy annular components.

After fitting the stiffness curves, the stiffness of Al6061 can be found, and shows a trend of increasing stiffness with increasing frequency for both unidirectional vibrations and two-dimensional coupling vibrations. Under low frequency vibration conditions, the deformation period of metal grains is much longer than the metal straining delay time, such that the internal stress within the metal grains has plenty of time to respond. Increasing the frequency of vibration, it was found that the deformation period of the metal grains gradually approaches, and may even be shorter than, the metal straining delay time. Therefore, the internal stress of the metal grains cannot be fully released.

It is impossible for the strain to reach the corresponding value at the same time of the stress. The deformation cannot establish the equilibrium state before the elastic wave is transmitted from the force application region and does not return from the object edge. Therefore, within the elastic limit loading strain behind the stress, but also need to overcome the static force between the grains, so the strain extremum of viscoelastic damping material under high-frequency simple harmonic load is smaller than the same amplitude of low-frequency excitation force, resulting in an increase of stiffness coefficient. At the same time, since the relative movements between the grain boundaries in the metal are affected by shear stress $\tau = G\frac{dv}{dt}$, the rate of variation of the vibration velocity v also increases with increasing vibration frequency. Hence, the shear stress must be overcome in order to also increase. As a result, stiffness of the metal is improved under high frequency vibrations.

After coupling the excitation vibrations in two directions, the value of the stiffness coefficient of the metal shows a significant decline within the high frequency range. This can be explained by the change in relative movement between the grain boundaries in the crystalline metal due to the coupling vibration. Under the effects of coupling vibrations, the vibration of the crystal in one direction can promote relative movement of the grain boundary in the same direction to cause sliding of the grain boundary. Moreover, the interaction forces between grain boundaries in another direction are weakened, thus reducing the required minimum force to initiate mutual movement between the grain boundaries. Under the condition of mutual coupling in multiple directions, it becomes easier to produce relative sliding between crystals, thus leading to a decrease in the stiffness of the metal material under the vibration coupling condition.

It should be noted, the internal microstructure of the component varies constantly under the action of high frequency loading [9]. Photos of the metallographic microstructure of the Al6061 test object used in the experiment are shown in figure 7.

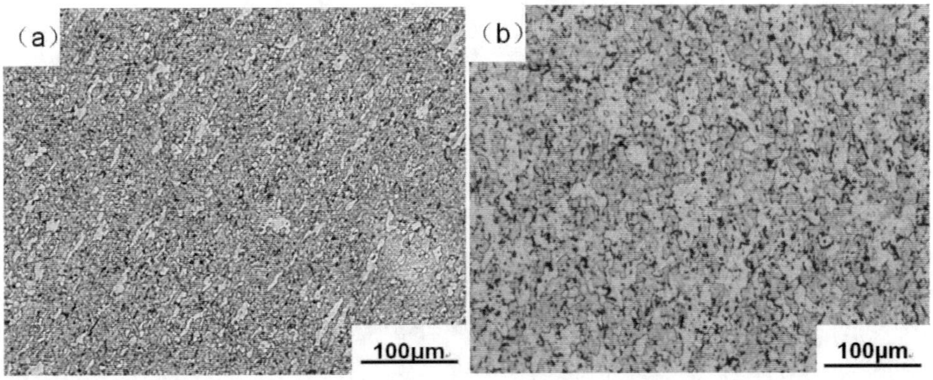

Figure 7. Microscopic images of the metallographic structure the Al6061 test object (40 mm external diameter; 2.5 mm thickness) used in the experiment. (a) Before experiment (b) After experiment.

From the comparative analysis of microstructures (Figure 7), it can be seen that prior to testing, the grains presented a clearly zonal arrangement with an average grain diameter of approximately 10.72 μm, in which black dots correspond to the strengthening phase of Mg_2Si (with a diameter of 5.14 μm), as shown in Figure 7 (a). After testing, the resultant microstructure is composed of equiaxed crystals, and grains have grown to approximately 19.31 μm due to thermal effects, and the grain diameter of the strengthening phase of Mg_2Si is 5.26 μm, as shown in Figure 7 (b). During the vibration process, induced obvious grain changes and grain boundary migration of the continuous shearing force.

Furthermore, in the experimental process some friction effects can be observed, as well as convective heat transfer between the shaft sleeve and test components, which further promote grain changes and grain boundary migration activities. Moreover, free grains can rotate by themselves, which leads to a constantly changing direction of thermal diffusion. Thus, under the comprehensive effects of various factors, the direction of grain growth varies constantly, and grains are gradually transformed into equiaxed grains. To verify the above conclusions, the thickness of the test object was increased to 4.5 mm, and the experimental results were found to be similar to those obtained for a thickness of 2.5 mm.

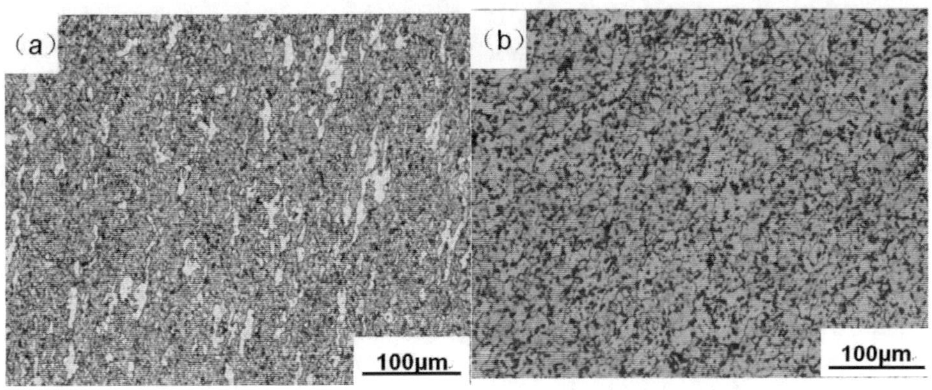

Figure 8. Microscopic images of the metallographic structure the Al6061 test object (40 mm external diameter; 4.5 mm thickness) used in the experiment. (a) Before experiment (b) After experiment.

IOP Conf. Series: Journal of Physics: Conf. Series **1183** (2019) 012013 doi:10.1088/1742-6596/1183/1/012013

Finally, if we change the material of the test object to hard nylon with a 40 mm external diameter and 4.5 mm thickness, changes in microstructure before and after experiment present a similar trend.

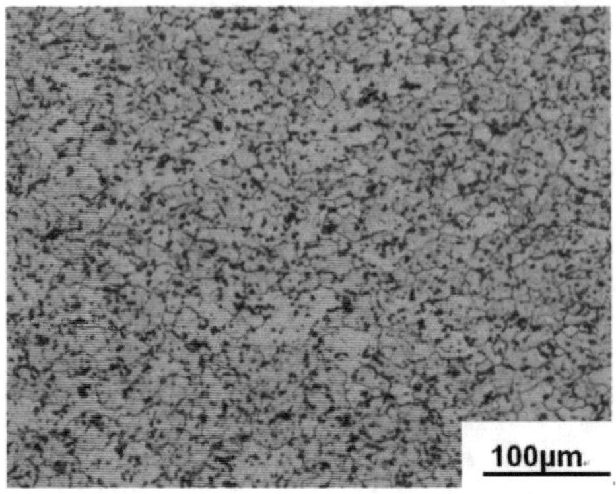

Figure 9. Microscopic images of the metallographic structure the hard nylon test object (40 mm external diameter; 4.5 mm thickness) used in the experiment.

4. Conclusions

In this paper, a system for testing the dynamic characteristics of rotor components was designed based on the BERM method, and the relationships between the different parameters (stiffness value, damping coefficient, and loss factor) of various damping materials and frequency was investigated. The stiffness coefficient of typical rotor parts was found to increase up to a frequency of 2000 Hz, and demonstrated typical nonlinear characteristics. If the rotor components suffered synchronal action from simultaneous axial and radial loading in two directions, the stiffness coefficient values were affected, and presented a decreasing trend. The main reasons can be attributed to the influence of the operational mode and microstructure of the internal grains. Based on the above conclusions, optimization can be achieved in the design of rotor components, and reasonable operating parameters can be set, thus enhancing the reliability of the system as a whole.

Acknowledgement

Supported by the National Nature Science Foundation of China (Grant No. 51205027): The optimization research on dynamic characteristics of the rotor part under the condition of superhigh stable acceleration.

References

[1] Department of Engineering and Materials Science (2010), National Natural Science Foundation of China. Mechanical Engineering Discipline Development Strategy Report. Beijing: Science Press.

[2] Atwell A R,Okojie R S and Kornegay K T (2003). Simulation fabrication and testing of bulk micromachined6H-SiC high-g piezoresistive accelerometers [J]. *Sensors and Actuators: A Physical*, 104(1) pp11-18.

[3] He Zongying, Jie Fengjuan, Guo Min, Li Minsheng (2010). Accelerated Testing under High Energy Shock Conditions [J] *Journal of Artillery Launch and Control*, 30 (3) pp329-333.

[4] Shi Yunbo, Zhu Zhengqiang, Liu Xiaopeng, Du Kang, Liu Jun (2010). Design and impact characteristics of high g value accelerometer [J]. *Explosion and Impact*, (1) pp85-88.

[5] Wang Tao, Wang Xiaodong, Wang Liding (2010). High g value acceleration environment microstructure dynamic characteristics [J] *Journal of Mechanical Engineering*, 48 (16) pp88-94.

[6] X L Lu, W Cai, L C Zhao (2003). Damping behavior of a $Ti_{44}Ni_{47}Nb_9$ shape memory alloy [J]. *Journal of materials science* letters 22 pp1243-1245.

[7] Justin D Marshall, Finley A Charney (2010). A hybrid passive control device for steel structures, I: Development and analysis [J]. Journal of constructional steel research. (66) pp1278-1286.

IOP Conf. Series: Journal of Physics: Conf. Series **1183** (2019) 012013 doi:10.1088/1742-6596/1183/1/012013

[8] Li Guoyong, Gu Zhengqi, Zhang Sha, Ma Xiaojiang, Li Jiping, Hu Kai (2015). Fatigue Life Prediction Method Based on Unequal Spacing Gray Model [J/OL]. *China Mechanical Engineering*, 26 (24) pp3267-3274.

[9] Chinh-Chun Hsieh, Chien-Hong Lai, Weite Wu (2013). Effect of vibration on microstructures and mechanical properties of 304 stainless steel GTA welds [J] *Metals and materials international*, 19 (4) pp835-844.

[10] Xuemei Luo, XiaoFei Zhu, Guangping Zhang (2014).Nanotwin-assisted grain growth in nanocrystalline gold films under cyclic loading[J]. *Nature Communications*, 3021(5). DOI:10.1038/ncomms4021.

Institute of Physics
Dirac House, Temple Back
Bristol BS1 6BE UK

ISSN: 1742-6588
ISBN 978-1-5108-8939-2

9 781510 889392

CURRAN ASSOCIATES INC.
proceedings.com